A Field Guide to Bacteria

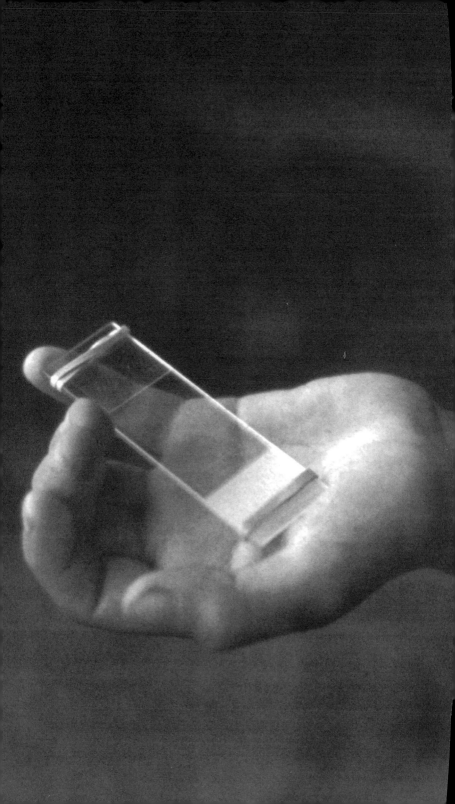

A Field Guide to Bacteria

Betsey Dexter Dyer

Comstock Publishing Associates
a division of
Cornell University Press

Ithaca and London

First published 2003 by Cornell University Press
First printing, Cornell Paperbacks, 2003

Printed in the United States of America

Library of Congress Cataloging-in-Publication Data

Dyer, Betsey Dexter.
    A field guide to bacteria / Betsey Dexter Dyer.
        p. cm.
    Includes bibliographical references and index.
    ISBN 0-8014-3902-7 (cloth)—ISBN 0-8014-8854-0 (paper)
    1. Bacteria—Identification—Handbooks, manuals, etc.
2. Microbial ecology—Handbooks, manuals, etc.   I. Title.

QR100 .D946 2003
579.3—dc21
                                                    2002041012

Cornell University Press strives to use environmentally
responsible suppliers and materials to the fullest
extent possible in the publishing of its books. Such
materials include vegetable-based, low-VOC inks and
acid-free papers that are recycled, totally chlorine-free,
or partly composed of nonwood fibers. For further
information, visit our website at
www.cornellpress.cornell.edu.

Cloth printing   10 9 8 7 6 5 4 3 2 1
Paperback printing   10 9 8 7 6 5 4 3 2 1

Contents

Contents

The color plates follow page 182

Acknowledgments

I was introduced to the microbial world by my mentor and Ph.D. advisor, Lynn Margulis. In 1980 she organized a course in planetary biology and microbial ecology (funded by NASA) which, in retrospect, I see as a critical milestone in my career as a scientist. It was during one of the many wonderful field trips to microbial hot spots, led by Stjepko Golubic, Wolfgang Krumbein, Ken Nealson, and others, that I realized that a field guide such as this one could (and should) be written. I believe the epiphanic moment came when Stjepko took us from the microscopic view (via his little hand-held field microscope) to the macroscopic view (and odors) of scums, slimes, felts, and bubbles of gas.

I am fortunate to be a professor of biology at a small liberal arts college, Wheaton College, in Norton, Massachusetts, where the project of researching and writing this field guide was so generously indulged over the course of about ten years. Colleagues who supported and tolerated my writing and editing include Tim and Gloria Barker, whose beach house overlooks a glorious assemblage of multicolored microbes, Scott Shumway, who answered my botanical queries, and John Kricher, who has been my mentor since 1971. Mark LeBlanc (my coauthor in genomics) helped me to stay on course and focus optimistically on every deadline. Advice from the English, classics, and religion departments was provided by Michael Drout, Joel Relihan, and Jonathan Brumberg-Kraus, respectively. Martha Mitchell at Wheaton's interlibrary loan office tracked down and delivered hundreds of reference papers and books to me. The four-volume set *The Prokaryotes* has been my primary source. I am indebted to its editors, Albert Balows, H. G. Trüper, M. Dworkin, W. Harder, and K.-H. Schleifer, for making the information so acces-

Acknowledgments

sible. Wheaton's provost, Susanne Woods, provided funds for several essential field trips, including to Yellowstone National Park.

Kathy Rogers did much more than type the mostly handwritten manuscript. During much of the writing process she was my only reader, and she kept me going with her generous enthusiasm and encouraging comments about every chapter that crossed her desk. She sustained me through each deadline, often more confident than I that we would make it. Kathy and I both refer to this field guide as our book.

Peter Prescott at Cornell University Press has watched patiently and persistently over many aspects of this project. I am grateful to Russell L. Cuhel, Barbara Eaglesham, Stjepko Golubic, Matthew D. Kane, Edward Leadbetter, John Stoltz, Douglas P. Zook, and several anonymous reviewers for their corrections and clarifications. Any errors that might inadvertently remain are certainly mine alone. My copyeditor, Allison Aydelotte, applied her pencil to nearly every paragraph. Allison gets credit for coining the word *bacteriophilic*, which I hope will become popular. Louise E. Robbins made many suggestions for clarifying and unifying the manuscript.

My parents, Jean and E. Otis Dyer, provided an influential childhood setting, their farm in Rehoboth, Massachusetts. The farm features an iron bog, composting manure piles, methanogenic swamps, nodulated legumes, and several head of bovine fermentation vats, and their basement housed my first research laboratory (complete with microscope).

My husband (and occasional coauthor) Robert Obar and our children Alice Linnea Obar and Samuel Darwin Obar have had many a family vacation focused around a search for bacteria. I am grateful for their enthusiastic cooperation on every occasion that I insisted we stop to examine a bit of scum or slime. They suggested that we get T-shirts printed with: "I was at an eruption of Old Faithful but didn't see it because I was too busy staring down at the bacteria in the runoff water."

A Field Guide to Bacteria

Introduction
Becoming Bacteriocentric

I love studying and thinking about bacteria, but this was not always the case. When I was growing up, I was a serious amateur naturalist. Insects were my favorite organisms, and I planned to become an entomologist. I was also very fond of rocks, shells, birds, bird nests, feathers, skeletons, leaves, and flowers. I had many collections and little experiments set up in my research lab in the basement. When I was eleven, I was given a microscope for Christmas and became quite enthusiastic about the world of microorganisms—but not about bacteria. Tiny crustaceans, ciliates, nematodes, and miniature annelids were abundant in drops of pond water. However, bacteria were below the limits of easy detection with my microscope, and even if I had seen them, I suspect that the tiny dots and dashes—the typical forms of bacteria—would not have held my interest.

In college, I took many biology courses and completed a project involving microscopy during my senior year. However, I did not appreciate bacteria then and, to be honest, I was even intimidated by them and preferred not to think about them at all. Somehow I avoided taking any course in microbiology as an undergraduate.

In graduate school, I became more comfortable with bacteria, since they live in abundance in the hindguts of termites—a subject of my graduate research. This continues to be my favorite microbial symbiosis. Top-quality microscopes and the unusually large and active bacteria characteristic of the termite microbial community all contributed to making my first real experiences with bacteria positive ones.

At that point in my career, I was already working at a professional or near-professional level in biology. However, I was still under the impression that an appreciation for and understanding of bacteria could be achieved only after years of work. At no point did it seem to me that I could have studied bacteria as an amateur naturalist. Even as a professional biologist, I found them quite difficult.

All that changed for me in the summer of 1980 when I joined other scientists on a field trip to the microbial mat communities in Baja California, Mexico. There we marveled at felt-like blue-green bacterial mats, pink layers of scum, black sediments, red-tinted salt crystals, and bubbles rising to the surface of murky water, all accompanied by a distinct sulfur smell. I learned that many bacteria can be identified by *macroscopic* field marks—characteristics that anyone can see, smell, or sometimes even hear, without the use of a microscope. It occurred to me right then that it would be possible to write a field guide to the macroscopic characteristics of bacteria.

Although most people are aware that bacteria are all around us, few would guess that they produce such distinctive and accessible signs. Whether you're walking on the beach, visiting a zoo or aquarium, buying groceries, looking for fossils, drinking beer, traipsing through a swamp, or cleaning scum from a dripping outdoor faucet, you're surrounded by bacterial field marks. You don't need a laboratory or fancy equipment to find out what kind of bacteria are there—this guide will tell you how.

If you want to know what to look for at the zoo or on the seashore, turn to the next chapter, "Guide to Habitats," or look in the index. Or maybe you prefer to read about a particular group of bacteria and take off on a search for them; in that case, start with any chapter in the book. The following section of the introduction gives more detail about how the book is organized and how to start becoming bacteriocentric.

How to Use This Guide

The purpose of this guide is to add a bacterial dimension to nature studies and to make more obvious the most abundant and diverse organisms on Earth—bacteria. This guide presents all the major taxonomic groups of bacteria in a useable, accessible format for *amateur naturalists* who may or may not have access to a microscope. This approach to bacteria is untraditional but is one that should appeal to users of field guides for other unusual subjects: fungi, minerals, bird calls, and the like. The sorts of challenges encountered in guides to fungi, for example, are the same that I have dealt with here: the subjects are cryptic, easy to overlook, and challenging to interpret.

This field guide is intended to be carried into the field by serious amateur naturalists, biology teachers at all levels, and even some professional biologists who may appreciate the accessibility it affords to these otherwise obscure organisms. Indeed, many serious amateurs and professional biologists are aware of the abundance and diversity of bacteria yet have always considered them to be off limits for simple study or appreciation. Although microscopy is encouraged (and detailed instructions given for those who have access to microscopes), the major theme of this guide is the use of macroscopic field marks to identify nearly every major group of bacteria. Indeed, the macroscopic identifiers are especially valuable to those who take an ecological approach to nature studies. With the aid of this guide, for example, bacterial components of the carbon, nitrogen, and sulfur cycles may be identified in the field, along with their more obvious plant and animal cohorts.

This field guide is organized by chapter according to the major taxonomic groups of bacteria. A taxonomic, rather than ecological, organization was chosen because it better reflects the extraordinary diversity of bacteria, even within closely related groups. Each chapter covers a branch on the bacterial family tree that has been constructed based on changes in DNA sequences that have occurred over the last 4 billion years. Thus, the current classification system and the organization of this guide both reflect the evolution of bacterial groups.

Because many of these bacterial groups are intimately associated with specific environments such as hot springs or marine mud flats, each chapter includes an ecological/environmental focus to place the bacteria in context with their surroundings. Summary lists at the end of each chapter also organize the macroscopic field marks for quick identification. The next chapter gives ideas for planning field trips to explore assemblages of bacteria in their natural environments.

Sometimes you never see the actual organism you have set out to study. Instead, you see signs of its presence or its activities. Take birds, for example. Certainly they are among the most watchable and identifiable of subjects for study. However, different manifestations of birds can also serve as identifiers: bird nests, bird eggs, bird tracks, bird songs, and even a brief flash of a bird, seen at such a great distance or so fleetingly that it becomes a real challenge to make any identification.

Think of this field guide as a naturalist's approach to those tiny, cryptic organisms that are at once the most populous and the least visible of living things on Earth. There are approximately 5×10^{30} bacterial cells on Earth according to University of Georgia microbiologists William Whitman, David Coleman, and William Wiebe—who refer to bacteria as "the unseen majority." Another 10^{18} bacteria, probably dormant, are circulating in the atmosphere attached to the dust of airborne soils and sediments, according to Dale Griffin and his colleagues at the U.S. Geological Service. But even with the aid of a microscope, most bacteria appear as nearly featureless dots and dashes displaying only a few cryptic behaviors (if jiggling about can be considered a behavior).

What can be done to make bacteria more accessible? In this guide I suggest looking for the manifestations, or *field marks*, of bacteria, which can be surprisingly visible and obvious once you know the signs. In a sense, this field guide is about what birders call "jizz"— the collection of characteristics of a particular bird (along with its habitat) that in total allow a well-trained observer to identify a bird in a matter of seconds, even as the bird is disappearing into the underbrush. Bacterial jizz is for the most part about what bacteria are doing, often on a large, detectable scale: producing bubbles, slimes, and scums; exuding odors and flavors; showing a stunning array of pigmentations; and participating actively and often visibly in nutrient and mineral cycles. In fact, in some cases bacteria are the sole proprietors of whole sections of these cycles. Without bacteria, the ecological wheels would cease to turn. Because different types of metabolism are such a central part of what differentiates one group from another, I describe metabolic processes in individual chapters rather than in the introduction. For example, you will find a description of photosynthesis in chapter 1 and chemoautotrophy in chapter 7 (consult the index to locate other topics).

Each chapter tells you where to go and what to look for (and smell, taste, or touch) to identify a particular bacterial group. In some cases (for example, cyanobacteria, methanogens, and intestinal microbes), one can even listen to the bacteria as they produce popping, fizzing bubbles of gas. Each chapter also has sections on what you will see if you look under a microscope, and, if it's possible to do so, how to culture (grow) bacteria in this group. All that is asked of the reader is to be open minded. Take a few trips to so-called extreme environments, such as hot springs or salt flats. Poke around and look closely. Everywhere bacteria are making their presence known, producing field marks that can be interpreted by naturalists at all levels. It is the aim of this guide to make those field marks accessible, interpretable, and less mysterious—and in so doing, to reveal the wonderful diversity of the bacterial world.

In Defense of Bacteriocentricity

Anthropocentrism—that tendency to see the universe in terms of human values and human experiences—is generally frowned on in modern nature writing. Indeed it is considered a rather unsophisticated point of view. Even more so in scientific research, anthropocentrism is thought to reflect a great error in logic, leading to false interpretations and a failure to see the whole picture. Most scientists (or at least their reviewers) are quite conscientious about editing out any traces of anthropocentrism from reports of field and lab studies.

Nevertheless, teachers of biology (and their students) know very well how useful a vivid analogy from personal experience can be. Some of us secretly appreciate it when "objective" science is put, even briefly, into human terms for the purpose of making the difficult a little more accessible. Thus, throughout this field guide I occasionally lapse into presumptuous anthropocentrism for the purpose of creating memorable examples of an otherwise cryptic and often overlooked world. But what I really want to say here is that I've done something far more presumptuous—and it is something that I recommend you try yourself. My primary goal throughout has been "bacteriocentricity"—that is, to put myself in the place of bacteria, to try to experience the world as they experience it. I have tried to see myself as enormously large (as indeed I am—most organisms on Earth are microscopic) and strangely multicellular (most organisms on Earth are unicellular). My range of metabolism is quite limited—centering only around oxygen respiration—and it takes me years and years to reproduce, which after all is the sine qua non of all living things. How strange that I don't bud or divide with bacterial frequency and efficiency.

A bacteriocentric point of view is a useful one to cultivate, whether you are trying to observe and understand the bacterial world yourself or to teach about it to others. Teachers in particular might want to lead students through the mental exercise of "being a bacterium." For example: Why is that pink scum positioned just so beneath the green scum? What does it "want" or "need"? What are its interactions with the other bacteria of the community? In fact, such an exercise is similar to one used by professional microbiologists when they try to culture bacteria. It's called "thinking like a bacterium"—trying to guess what parts of the microenvironment are essential, what aspects might be combined in a test tube to create the right conditions. (It is the method described in appendix A, on culturing bacteria.) A microbiologist wonders if there exists a bacterium that prefers high temperatures, low-nutrient conditions, and an acidic environment—and accordingly tries to establish those conditions in the lab in the hopes of attracting and nurturing such a microbe. It works!

Therefore, reader, try seeing the microbial world from a microbial point of view. Risk the presumptions of bacteriocentricity for a better understanding of the most predominant and most invisible organisms on Earth. Can this approach be taken too far? Perhaps. While I was immersed in writing this guide, a friend asked me for advice concerning a mild case of food poisoning he was experiencing. I found that I was unable to properly sympathize with him (the human host) but instead came down quite strongly on the side of his intestinal bacteria, which, after all, were experiencing an invasion and were being dislodged from their habitat and deprived of their usual nutrients.

TAXONOMY AND NOMENCLATURE

Attempts to classify bacteria have been fraught with difficulties. Most bacteria are tiny rods, indistinguishable from each other; many of the rest are tiny spheres. No wonder bacteriologists have turned to other characteristics more prominent than morphology! Metabolism—the means by which organisms get energy and food—is quite diverse among the bacteria and has traditionally been the primary characteristic for distinguishing them.

Since the 1980s, DNA sequences have become the basis for classification. Techniques for sequencing DNA have advanced so far that the process is now routine and widespread. Certain sequences that are highly conserved, meaning that they have undergone few mutations, are good candidates for constructing phylogenetic trees of organisms. One such sequence consists of DNA that codes for the 16S subunit of ribosomes, which are ubiquitous cell structures that are used to synthesize protein. The 16S sequence, so-called because its weight and size cause it to sediment in a centrifuge with particles designated as 16 Svedbergs, is one of the most trusted sequences for constructing family trees. Carl Woese, at the University of Illinois, and Mitch Sogin, at the Marine Biological Laboratory in Wood's Hole, as well as many scientists associated with their labs, have been among the most prolific of researchers producing the 16S ribosomal sequences for bacteria and other organisms.

As more and more bacterial genes were sequenced during the 1980s and 1990s, there came some big surprises. First, bacteria, or prokaryotes, could be divided into two large, very different groups, now named the archaea (or archaebacteria) and the bacteria (or eubacteria) (fig. I.1). Second, metabolism, once a trusted indicator of bacterial phylogeny, was found to be an unreliable trait for most groups of bacteria. Most of the major bacterial groups presented a seeming jumble of metabolic diversity. Third, some groups once considered "primitive" now became positioned on advanced branches, and vice versa. The new bacterial phylogeny that emerged was at first a challenge to explain and defend. However, this new phy-

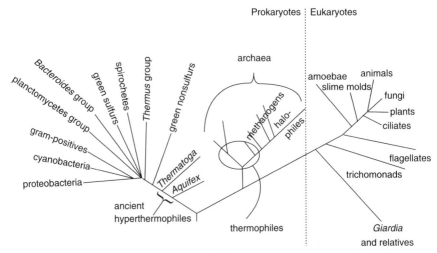

FIGURE I.1. This phylogenetic (family) tree shows the divergence of different groups of organisms from the beginning of life on Earth, about 4 billion years ago, as measured by differences in DNA sequences. The evolutionary distance between two groups is proportional to the total length from the tips of the two branches to the node that joins them. The thermophilic archaea include the hyperthermophilic species featured in chapter 3.

logeny also produced new ways of understanding bacterial evolution. Among other things, it appears that most of the great diversity of bacterial metabolism evolved early; recent branches merely display variations on several major themes.

Bacterial taxonomy based primarily on DNA sequences has now been incorporated into most publications concerning bacterial classification. In the long run, it may appear to historians of science that the breakthrough work of Woese and Sogin—and the almost complete rearrangement of our understanding of taxonomy and evolution—produced a true paradigm shift. The organization of this field guide is based on this new (and still shifting) taxonomy, allowing the reader to identify most major groups of bacteria, as well as many of the subgroups.

HOW MANY SPECIES OF BACTERIA ARE THERE?

The smaller or more inaccessible an organism is, the more difficult it is to answer the question, "How many species are there?" We know with considerable confidence how many mammals and birds inhabit the Earth. It is much more difficult, however, to know the numbers of flies or deep-sea worms. Most challenging of all are the bacteria. How many bacterial species are there? More than 2,000 have been described, some with great precision. The best-known of these—such as the gram-positives and the proteobacteria, which occupy several chapters of this field guide—can be cultured readily in the lab and have significant medical and economic importance to humans. We know what these bacteria can and cannot "eat" and what conditions favor their growth. The 300 to 500 or more genera are divided into 2,000+ species, with many of them among the famous "lab rats" such as *Escherichia, Streptococcus, Staphylococcus,* and *Pseudomonas.* When microbiology students are given the task of identifying an "unknown" bacterium, they are almost invariably assigned an easily cultivated species from among the gram-positives or the proteobacteria.

Has the number of bacterial species been *under*estimated? Many field microbiologists would answer this question with an emphatic "Yes!" Every year, we learn a little more about the microbes living at great ocean depths, in polar ice and boiling springs, and encrusted within deep sediments. New species of bacteria are continually being discovered and named, with no end in sight. Perhaps there are many thousands more to be added to the list—some scientists have estimated that less than 1 percent of bacterial species have been cultured and characterized. Perhaps, too, the characterizations of new species will not be as readily organized in the neat identification tables found in microbiology lab manuals. The fact that there are many more recorded species of gram-positives and proteobacteria than of other groups may reflect not their real-world abundance but rather the ease with which these groups can be studied and categorized in the lab.

Has the number of bacterial species been *over*estimated? Strangely enough, this question can also be answered in the affirmative because of the difficulty in firmly assigning bacteria to the level of species. The modern concept of species was first developed to apply to large, sexual organisms for which producing fertile offspring when mating outside of the species is, by definition, not possible. Bacteria do not reproduce sexually and have therefore never been assigned to species according to this definition, but rather according to differences in their culture characteristics in the lab.

But here is the problem: Bacteria are extraordinarily promiscuous with their DNA. By a process called "horizontal transfer," DNA sequences can be exchanged among different species and even among different kingdoms. Horizontal transfer of DNA is not only possible, but it is apparently carried out readily by intermingling bacteria in the wild. As a result, then, bacteria can acquire DNA sequences not only from each other but from humans, for example, effecting a transfer of DNA from the animal kingdom to the bacterial kingdom. Some microbiologists (such as Sorin Sonea and Maurice Panisset) have suggested that there are really no bacterial species at all but rather a sort of continuum of flowing genes over a huge amount of space and time. At any given point we have a snapshot that gives us the illusion of taxonomic groups because exchanges occur most easily between similar bacteria and less easily between more distantly related groups. Similarly, bacterial groups that are not physically near to each other cannot easily exchange DNA, whereas those that inhabit the same space may benefit from having interdependent metabolisms: two partners in a tight consortium may not both need to maintain a full set of genes.

Thus we should—with caution—use the analysis of DNA sequences to assign bacteria to large taxonomic groups but also realize that fine differences in sequence, metabolism, and morphology may not be reliable differentiators of genus and species. Does this mean that the concept of species should be abandoned for bacteria? No. Identification methods continue to be very useful for the diagnosis and treatment of pathogens. Furthermore, many lab techniques remain our only way of sorting out and understanding the largely invisible world of bacteria.

Much of what we know about bacterial taxonomy has been hard won by clever, creative researchers using methods that are well worth learning and practicing. But taxonomists are also cautious about applying this information too simplistically when assigning phylogenetic groups, especially genus and species.

NAMING BACTERIA

Serious naturalists and professionals in biology take pride in knowing names of phyla, orders, families, genera, and species. This process of assigning names is, of course, much more than just a mental exercise. It is essential because regional nicknames or common names for organisms are often ambiguous or limited in their usage; we need scientific terminology to communicate outside of our own locales. However, while this field guide is full of Latin- and Greek-derived names, it is also full of common names. This requires some explanation.

First, in ordinary conversations about their organisms, professional microbiologists often use nicknames (some, seemingly affectionate), and at the very least use truncations of much longer terminology—referring, for example, to the alpha proteobacteria as "alphas," the Chlorobiaceae as "green sulfurs," and the methanogens as "methanos." (This nicknaming is not obvious in the professional literature, which is almost always written in strictly formal style, but, in microbiology in particular, perhaps it helps researchers feel a little closer to their subjects, making the invisible less so.) I've attempted throughout to use the common names known to many professional microbiologists, in the hopes of letting the reader in on the sometimes casual nature of microbiological field and lab talk.

In spite of nicknaming, microbiologists take their scientific terminology very seriously. The seriousness takes two forms:

1. Considerable careful and persistent effort is made to identify bacteria to the level of genus and species. This task is not easy in most cases. A microbiology lab can look like a chemistry lab, with numerous test broths and multicolored stains. In fact, most lab training of microbiologists centers around the many techniques used for identifications. In most cases, however, it is well beyond the purview of this field guide to lead the reader through identifications to genus and species. The few exceptions are those bacteria that display extraordinarily specific field marks.
2. Bacterial phylogeny has been in a welcome state of flux, with even phylum-level classification being overturned based on new evidence from DNA. However, many bacterial groups still remain ciphers. Therefore, microbiologists generally try to show caution and restraint in coining new names. Until we know exactly what a bacterial group is and does and where it properly belongs on the family tree, we use relatively noncommittal names. A good example can be found in the proteobacteria, in which

the new subclassifications based on DNA evidence have simply been assigned the Greek letters alpha through epsilon instead of elaborate (and often multisyllabic) Greek or Latin names.

Most current bacterial terminology plunges from the level of phylum (or something like phylum) to genus and species, bypassing completely any mention of order. For example, *Bergey's Manual of Systematic Bacteriology* (1994) places green sulfurs in major category I (gram-negatives), group 10 (the anoxic phototrophic bacteria), and subgroup 4, whereas *The Prokaryotes* (1992) places them in the sole family in division E (both family and division called Chlorobiaceae). However, most bacterial family trees based on DNA sequencing assign the green sulfurs their own major branch, suggesting that they are, at the least, a phylum. Because of this lack of consensus, it is often not possible in this guide to identify what level (for example, phylum, class, or order) a bacterial group such as green nonsulfurs, spirochetes, or gamma proteobacteria corresponds to.

Therefore, I ask that you bear with the mixture of seemingly dignified Greek and Latin names interspersed with unmemorable temporary assignments, such as "alpha proteobacteria," seeing them instead, perhaps, as reflections of these exciting, unsettled times in bacterial taxonomy—for professionals and amateurs alike.

Balance and Focus

This guide presents all the major taxonomic groups and many of the subgroups of bacteria. With the exception of planctomycetes, which are inferred rather than identified, all the major groups are identifiable in the field without a microscope. Nevertheless, balance may be of concern to some readers. Have any important bacteria been left out by an emphasis on those bacteria with the most rewarding, most accessible macroscopic field marks? Indeed, this guide presents not a balanced treatment but rather one that is intended to be inviting to those who would not otherwise consider looking for or at bacteria. Many bacteria presented in traditional microbiology classes—many of these well-studied pathogens—are treated very briefly because they are identifiable only in professional labs.

Actually, although it is not always acknowledged, balance is a problem for any book on bacteria. For example, many microbiology textbooks (and courses) are dominated by a few groups of medical importance, often leaving the impression that most bacteria are pathogens. Textbooks on environmental microbiology often manage a more even presentation, but nevertheless do not do as much as they could with macroscopic indicators. Much of the information on quick macroscopic field identification gets passed along through word of mouth (from expert to expert) or is published in technical papers. Even *Bergey's Manual of Systematic Microbiology*, long considered the bible of the microbiology lab, devotes much more space to well-studied groups—often those bacteria that proliferate like weeds in the lab—than to the more fastidious or slow-growing species that actually predominate in the environment.

Professional microbiologists will be quick to discover that the sections pertaining to their own particular bacteria of expertise are by no means exhaustive—nor are they intended to be in the usual professional context. Rather, this is a collection of tricks of the trade and rules of thumb by which professionals identify bacteria in the field. Many of the methods are communicated by word of mouth or in the medieval guild style of passing information from master to apprentice, still typical of many Ph.D. programs. Perhaps professionals will appreciate a written compilation of their "trade secrets" for macroscopic identification of bacteria.

So, yes, this book too has a slant. It tilts in the direction of those bacteria that are most likely to be identified in the field, without resort to the usual battery of laboratory tests and extensive microscopy. It is a happy coincidence for the organization of this guide that so many of the major taxonomic groups and subgroups are identifiable directly through the senses. Or it might be no coincidence; after all, bacteria are the most numerous and diverse organisms on Earth. Their apparent invisibility may be just an artifact of a human-centered point-of-view.

ECOLOGY ON A MINIATURE SCALE

INTIMACY WITH THE ENVIRONMENT

"Microscopic" is the normal scale of the world. Most organisms on Earth are tiny and always have been and probably always will be.

The relatively huge, lumbering size of ourselves and our fellow animals is exceptional—albeit quite successful in its own way, especially if you use beetles as the representative animal. J. B. S. Haldane, a mathematical biologist, noting that there are 8,000 species of mammals but 400,000 species of beetles, commented that the creator must have had "an inordinate fondness for beetles."

What does it mean to be microscopically small? Smallness means to be in intimate contact with the environment. A little moisture or dryness, slightly more concentrated salt, an elevated temperature or pH are all sensed directly by single cells. The responses produced are rapid, usually involving some change in metabolism, even shutting down to dormancy if conditions are too stressful.

Like all organisms, bacteria need water to live; microbial ecology is aquatic ecology. On the microscopic scale, "water" could mean a droplet on a leaf or a moist soil particle, the interior of an animal's intestine, the open sea, or the hot sulfury waters of a thermal spring. Water—in some form—is needed for cells to be active and to reproduce.

While ecology on a larger scale is usually divided into terrestrial and aquatic habitats, microbes experience the environment as an aquatic continuum. The moist soil at the side of the river is continuous with the river water and the water that has splashed up on the rocks. The moist intestinal environment is a continuum with the moist feces and moist soil on which the feces are deposited. Similarly, the dry desert is merely an aquatic environment in which the dormant inhabitants are waiting for rain, at the dry end of a temporal continuum.

Rather than terrestrial and aquatic, the microbial world is divided according to what nutrients and energy sources are available and what chemical and temperature conditions are present. Such parameters may change dramatically (from a microbial point of view) over the length of just a few bacterial cells. The bacterial world can be thought of as a mosaic of tiny islands of discontinuous resources.

Life on a soil particle, for example, is quite different from life on the surface of a nearby rootlet of a plant. Life in the top layer of a microbial mat is different from life even a few millimeters down. Understanding what bacteria are taking from their environment and depositing as waste is the key to understanding microbial ecology and the distribution of microbial species, as well as their identification in the field.

MOVEMENT

Being tiny means experiencing almost any aquatic environment as extremely viscous. This phenomenon was discovered and quantified in the early 20th century by an engineer named Osborne Reynolds. He came up with a measure of fluid flow called the Reynolds number, which helps to understand the relationship between size and the ability to glide through a watery environment of a particular viscosity. A whale in the ocean has a huge Reynolds number (about 300,000,000), which means that with a few strokes of its fins it can glide a great distance. Humans glide less efficiently because they are smaller and have a Reynolds number of about 10,000. Applying this engineering concept to bacteria was the brilliant idea of Harvard biologist Howard Berg. A bacterium in the ocean does not glide at all. If it stops twirling its flagellum, it comes to an immediate stop. It has a Reynolds number of about 0.00001. You can experience a similar phenomenon by trying to swim in a pool full of gelatin (or perhaps just use your imagination). Even if you push off strongly from the side of the pool, you (bacteria-like) will fail to glide through the viscous gel. That's why so many bacteria don't even "try" to glide. Instead they attach to a surface and allow the watery environment to flow by. Motion is relative—flow brings nutrients and removes waste, and that is often all a bacterium needs. Rapid flow can be really inconvenient for a tiny bacterium, however, if it wants to maintain its position; that's why the bacteria in rivers and intestines so often live firmly attached to surfaces. Attachment can also mean clinging to surface films of still water, to minute particles, and of course to other organisms. Another strategy is to float, or to at least maintain some approximate position in the water by means of gas-filled vacuoles.

Motile bacteria that use tufts of flagella or gliding mechanisms rarely move far. Usually they jockey for position in a community, finding optimal positions in the sun or in a gradient of sulfide or oxygen or temperature. A few millimeters is a long way. The major exception to this rule can be found in the spirochetes (chapter 16)—

long, skinny organisms so efficient at motility that they seem to defy the restraints of the Reynolds numbers. They can move through the most viscous of environments, such as thick mud or full intestines. This means of motility is one reason why some of them can be such insidious pathogens, corkscrewing through our tissues.

SIZE

There must be something especially advantageous about being tiny. The logic is simple: Most organisms are tiny bacteria, and the greatest organismal diversity is found among these tiny bacteria. Furthermore, most of the evolutionary history of Earth has been dominated by (tiny!) bacteria. What is so wonderful about that size? Bacteria have a certain intimacy or immediacy with the environment such that slight changes in a local area can produce rapid and specific responses. In addition, reproduction (cell division) in bacteria is faster than it is for larger cells or for multicellular organisms. Perhaps that kind of efficiency counts for a lot. There seems to have been little deviation through evolutionary history from the fast-responding, rapidly dividing body plan typical of most bacteria. Therefore, the truly large bacteria are worth speculating about: Why big?

Only a few bacterial groups have gigantic representatives: the gram-positives, the proteobacteria, and the spirochetes. These are also three of the most thoroughly studied groups of bacteria. Cyanobacteria should also be mentioned here, although they are for the most part large and conspicuous due to multicellularity.

How gigantic is "gigantic"? The behemoth bacterium *Epulopiscium fishelsoni*, a gram-positive species that inhabits surgeonfish guts, is a strong candidate for the largest. It is a rod about 80×600 micrometers in size—in other words, 0.08×0.6 millimeters, or about half a millimeter long, bigger than most animal cells. If you are willing to squint at specks, you can actually see this cell with your naked eye. Which leads us to ask, is there anything special about surgeonfish guts? These fish eat algae, which puts them in the same league with terrestrial herbivores such as cows. Plant and algal material can be difficult to digest, and all known herbivores cultivate symbiotic bacterial communities in some part of their digestive system to help with that process. Guts are safe, nutrient-rich places, extremely popular as habitats especially among fermenters such as gram-positives. Some of the largest bacteria I have ever seen—long gram-positive rods—were in the symbiotic community of termite guts. The abundance of nutrients in such a habitat might allow bacteria the luxury of time and growth. Therefore, while we might not expect to find truly gigantic bacteria in every herbivore gut, we might expect to find many large ones.

Speaking of guts, that is where we might expect to find some of the largest spirochetes too (see chapter 16). Again, it has been my experience in looking at termite digestive symbionts that the spirochetes are especially large (up to 100 micrometers long) and active. Also, the spirochete *Cristispira*, found in some molluscan digestive systems, can be up to 150 micrometers long (but only about $1^1/_2$ micrometers wide). You still need a microscope to see them. The biggest of the spirochetes—*Spirochete plicatilis*—may be free living, enjoying the rich environment in and around strands of *Beggiatoa* (see chapter 9) in some sulfur-rich environments. If you are looking at *Beggiatoa* (highly recommended) you may find this spirochete too. It can be up to 250 micrometers long (a quarter of a millimeter) but less than a micrometer thick.

Among the proteobacteria, the large bacteria, such as *Beggiatoa*, are those that perform a fairly delicate and unusual metabolic procedure: sulfide oxidation. Their size may be related to their metabolism. This process requires having a source of unoxidized (reduced) sulfur, such as hydrogen sulfide (the smelly gas of salt marshes), along with a supply of oxygen. Oxygen, however, readily oxidizes hydrogen sulfide without any assistance from or benefit to the bacteria. Perhaps in some of these bacteria, large cells provide a way of keeping reactive compounds segregated. *Beggiatoa gigantea*, for example, can have cells up to 25 to 55 micrometers wide and about 10 micrometers long, arranged in long multicellular filaments. Two species of *Beggiatoa* from deep-sea thermal and cold-seep communities have cell diameters of 120 to 200 micrometers. These cells actually have relatively little cytoplasm (cell contents), as most of the space is a large storage vacuole. In these bacteria, nitrate is probably used to oxidize sulfide rather than oxygen, and it may be that nitrate is stored in the vacuoles.

Another strong contender for the "biggest bacterium" prize—

Thiomargarita namibiensis (sulfur pearl of Namibia)—is also known to store nitrate. This species consists of a balloon-like cell typically 100 to 300 micrometers but sometimes as much as 750 micrometers, or $3/4$ of a millimeter, in diameter. Look at a metric ruler and imagine a pearly droplet that just fits between two of the millimeter lines. Whether it truly deserves to take the prize from *Epulopiscium* (the species found in surgeonfish guts) depends on how you want to count the enormous nitrate-filled vacuole of *Thiomargarita* that comprises 98% of the cell. *Thiomargarita* has been described by microbial ecologist Bo Jørgensen as "holding its breath" by stock-piling so much nitrate.

The cyanobacteria get lots of credit for being very long, albeit very skinny (less than a micrometer). If you gently pull apart the top blue-green layer of a microbial mat and look closely, you will see the tiny lint-like strands that hold the layer together, like felt. These minute strands are multicellular clumps of long filamentous cyanobacteria, not individual cells. Long filaments may consist of hundreds of cells. Thus, most cyanobacteria are not truly among the largest bacteria. However, *Oscillatoria princeps* is of interest because its cells are quite thick (up to 60 micrometers).

ACTIVE VERSUS DORMANT SPECIES

A critical problem in microbial ecology is knowing which bacteria are normally active in a particular habitat and which are primarily dormant. The distinction can be subtle. Active bacteria use whatever nutrients and other conditions are supplied by the surrounding habitat. If resources are typically scarce, then the active population is likely to have a slow growth rate and may be difficult to detect. In contrast, the dormant population is inactive, either in a resistant spore form or some other quiescent state, until there is an influx of unusual nutrients. A new and transient source of rich food, not typical of the usual available resources, such as a dead beetle, may allow some dormant bacteria to flourish temporarily. It matters little where the beetle died, as a variety of dormant bacteria are almost always present in nearly all environments, resting until an opportunity occurs. The microbiologist M. J. Carr called this "watchful waiting."

Many dormant bacteria are also called "opportunists"; opportunists are among the easiest bacteria to detect, and they include a number of pathogens. Simply put a soil particle on some rich nutrient medium, and opportunists will be the first to grow. The more typical active population of the soil, in contrast, may never get a chance to demonstrate its slow reproductive rate. Therefore, routine microbiological culture methods are not always the most accurate way to detect and evaluate typical (often low-level) bacterial activity in an environment.

PATHOGENS

The pathogenic bacteria (sometimes referred to as germs) get plenty of publicity in spite of being in the minority among microbes. This is perhaps no surprise: we are human-centered creatures, concerned and even obsessed with any organisms that might do harm to us, our pets, or our food supply. Relatively forgotten or ignored are the myriad microscopic populations of soils and waters, most of which have no interest whatsoever in inhabiting or injuring the human body. Many bacteria are photosynthesizers (makers of their own food using light energy) or chemosynthesizers (makers of their own food using chemical energy). Other bacteria would never survive in the temperate environment of the human body, preferring instead much greater extremes of temperature and salinity. Others are quite conservative in the range of nutrients that they use, and would find the human body's rich and diverse mixture of chemicals a hostile environment. And for the bacteria that *do* live in and on us, there is, in most cases, no particular advantage to gravely injuring or killing their host. Much better is a coexistence of host and bacterium.

Furthermore, the human body (like any animal, plant, or fungus) is extremely well defended. We have a tough, dry, inhospitable skin and a vicious immune system which, when functioning properly, defends against the majority of potential invaders. Not only is it vicious, but it is also sometimes too easily aroused. In *Lives of a Cell*, Lewis Thomas described having an immune system as being surrounded by a minefield—one of danger to ourselves as well as to bacteria. Overstimulation of the system can result in fever, tissue damage, and shock. Mild stimulation brings on allergic reactions of all sorts as well as autoimmune diseases. Indeed, bacteria are not solely responsible for all of the detrimental or even lethal outcomes that can occur from infections; a massive inflammatory response can be just as harmful.

We also have our own indigenous—and often helpful—population of bacteria covering our skin and lining our digestive system, making it difficult for other bacteria to enter the body. A moderate exposure to bacteria of all kinds may prime the immune system to withstand later invasions. This is not to say we should be overly casual. Some bacteria arrive like the Trojan horse, tricking the immune system with chemical disguises (although this is, fortunately, relatively rare). Others get through breaches (cuts, scratches, bites) in the skin or digestive lining and reproduce quickly in places not easily monitored by the immune system. Some of our own indigenous bacteria, with which we normally have a symbiotic relationship, can turn nasty if there is a change in our usual defenses—such as a compromised immune system. Symbiosis is often not so much about mutual altruism as about mutual control. As Lewis Thomas put it in *Lives of a Cell*, "Disease usually results from inconclusive negotiations for symbiosis, an overstepping of the line by one side or the other, a biological misinterpretation of borders."

Who are the pathogens? They are a minority of bacteria. They tend to be opportunists, capable of reduced activity and even dormancy until a rich source of nutrients (such as ourselves) should suddenly appear undefended. Then they escalate their activities, consuming a variety of nutrients and multiplying quickly. Most of the best-known pathogens are members of the gram-positives and the proteobacteria, and others are spirochetes. *Chlamydia* species (relatives of planctomycetes) are all obligate inhabitants of other organisms' cells. The virulent members of these groups are well studied, well publicized, and, at least in industrial countries, fairly well controlled; several, however, are top killers in developing countries (see table I.1). Most of the pathogenic bacteria are not covered in this field guide—which would otherwise become in part a medical manual—but especially well-known pathogens are mentioned in passing when nonpathogenic bacteria of that group are discussed. Indeed, one of the more noteworthy characteristics of the pathogens, especially from the point of view of this guide, is just how limited they are to certain taxonomic groups and how rare their activities are within those groups.

TABLE I.1. Diseases of Humans and Other Animals, and the Bacteria That Cause Them

Disease or condition	Causal group and genus			
	Proteobacteria*	Gram-positives	Spirochetes	Chlamydia
Anthrax		Bacillus		
Cholera	Vibrio (γ)			
Diphtheria		Corynebacterium		
Ear infections	Haemophilus (γ)	Streptococcus		
Gastrointestinal infections** (including food poisoning)	Salmonella (γ) Shigella (γ) Campylobacter (ε) Escherichia (E. coli) (γ)	Clostridium (botulism) Staphylococcus		
Leprosy		Mycobacterium		
Lyme disease			Borrelia	
Meningitis**	Neisseria (β)	Streptococcus		
Plague	Yersinia (γ)			
Pneumonia**	Legionella (γ)	Streptococcus Mycoplasma		
Rocky Mountain spotted fever	Rickettsia (α)			
Sinus infections	Hemophilus (γ)	Streptococcus		
"Staph" infections		Staphylococcus		
Stomach ulcers	Helicobacter (ε)			
"Strep" throat (scarlet fever)		Streptococcus		
Tetanus		Clostridium		
Tuberculosis		Mycobacterium		
Typhoid	Salmonella (γ)			
Typhus	Rickettsia (δ)			
Venereal diseases	Neisseria (gonorrhea) (β)		Treponema (syphilis)	Chlamydia
Whooping cough	Bordetella (β)			

* Greek letter indicates alpha (α), beta (β), gamma (γ), delta (δ), or epsilon (ε) group.
** There are also viral versions of this disease.

OBSERVING BACTERIA

HOW EXTREME IS YOUR BACKYARD?

How many different bacterial groups can you see from your backyard? The answer to this question depends on how extreme your backyard is. Most people do their landscaping with an emphasis on eukaryotes (mainly plants and animals). A rule of thumb is, the more visible the eukaryotes, the more invisible the prokaryotes (although the latter are present by the trillions, far outnumbering animals and plants). Bacterial field marks are not so obvious against an overwhelming background of eukaryotes, especially as observed from the point of view of a fellow eukaryote.

Another rule of thumb is, the greater the diversity of species—whether prokaryotic or eukaryotic—supported by a particular environment, the less likely it is that any individual prokaryotic species will predominate. In other words, the more nonspecific and moderate the environment, the more likely there are to be many different types of prokaryotes, with each type in relatively low numbers. Moderate environments include ordinary marine and fresh waters and any soil supporting a diversity of plants. Environments featuring some extreme condition, such as high or low temperatures, acidity or alkalinity, high salinity, or a lack of oxygen, are likely to appeal to only a few types of organisms—and typically these organisms are prokaryotes. It is often in such extreme environments that prokaryotes predominate to the point of displaying prominent field marks.

Viewing a wide variety of prokaryotes, then, involves traveling to a wide variety of habitats. Because most prokaryotic groups are distributed worldwide, the same types may be found wherever similar habitats exist. Salt flats in Mexico have most of the same bacterial types as salt flats in Utah or in Israel. Hot springs in Japan have most of the same bacterial forms as hot springs in Wyoming and Iceland. Thus, you will have to travel out of your backyard, but not necessarily out of your country, to see all of the bacterial groups.

WHAT WILL YOU MISS BY NOT LOOKING THROUGH A MICROSCOPE?

Most bacteria are tiny rods or spheres, almost indistinguishable from each other. Usually what makes a bacterial species taxonomically distinct is what it *does*, not what it looks like. Bacteria take in various chemicals from their environments and put out other chemicals as products or waste products. In large enough numbers, they can produce substances that can be seen (or smelled or touched or tasted or even heard) on a macroscopic level. These field marks, along with characteristics of the habitat, are the basis for bacterial identification in this field guide.

I love microscopy, though, and I do not wish to make light of how exciting and important a tool it is for bacteriology. Some bacteria are quite distinctive in their morphology. Many photosynthesizers, for example, are large and colorful; helical spirochetes spiral easily through dense substrates. If you do not use a microscope, you are indeed missing out on the experience of seeing these bacteria directly. Therefore, consider using this guide as a starting point in your search for bacteria. This book will tell you how to find abundant, identifiable populations of bacteria that you can later scrutinize up close with a microscope.

Keep in mind that you will need a fairly powerful microscope, one with a capability of magnifying 400 to 1,000 times. Bacteria are about 1/10 to 1/100 the size of the more common microorganisms such as *Paramecium* and *Amoeba* that you may have encountered in an introductory biology lab. Without proper lighting and good lenses, bacteria can look much less distinct than tiny dots and dashes—they can appear to be nothing at all!

Throughout this guide, you will find sections on viewing a particular group under the microscope; these provide hints on what to look for if you have the opportunity to collect a sample and observe it. Appendix C also gives specific instructions on how to use a microscope to observe bacteria.

Guide to Habitats

The organization of this guide generally follows taxonomic order. However, bacteria, especially those with macroscopic field marks, are often closely identified with particular ecological communities. One way to use this guide is to go into the field seeking not just a particular taxonomic group but a whole assemblage of bacteria from different groups, all living and interacting together in a specific habitat. This guide to habitats provides suggestions for organizing field trips to places such as the seashore, a temperate forest, a farm, or a gourmet food shop. Some field trips, such as a visit to Yellowstone National Park, are incorporated into specific chapters. Others may be planned by using the index. Field trips can also be organized around adding to an insect, shell, or other collection. An advanced theme for a field trip is to seek out parts of a particular biogeochemical cycle. Finally, you can "plan" imaginary trips, such as a search for bacteria-like extraterrestrials on other planets.

FIELD TRIPS INCORPORATED INTO SPECIFIC CHAPTERS

HOT SPRINGS
Chapters 1, 3, and 17 are devoted entirely to bacteria of hot springs. Some cyanobacteria can also live in hot water; therefore, read chapter 13 for characteristics by which cyanobacteria are identified. Keep in mind that the runoff from hot springs cools and then reaches ambient temperature. Therefore, hot springs may provide good views of temperate bacteria. Furthermore, hot springs may be sulfur-rich or iron-rich. Therefore, check chapter 9 on sulfureta and chapter 7 on iron-oxidizing bacteria. Note too that chapter 3 is specifically about boiling, acidic, sulfurous springs.

SALT FLATS AND OTHER HYPERSALINE ENVIRONMENTS

Chapter 4 is focused on the salt-loving bacteria. See also descriptions of cyanobacteria in chapter 13, as some of these are salt tolerant. Salt flats are often rich in sulfur compounds. Therefore, plan to read chapter 9 on sulfureta for a more complete picture.

SULFUR-RICH ENVIRONMENTS

The full panoply of sulfur-based bacterial activities is highly accessible and often quite colorful and interesting. If you can smell sulfur with or without stirring up the water or sediment, it's probably a sulfur-rich environment (see chapters 5 and 9). Such environments include marine waters, such as intertidal flats and estuaries, and fresh waters, especially in areas with sulfur-rich sediments, as well as sulfur springs. Polluted (or overfertilized) soils and waters may also smell sulfury. If you visit boiling sulfur springs, consult chapter 3. Keep in mind too that some salinas (high salt) areas are sulfury. As with many of these field trips, it is worthwhile to familiarize yourself with cyanobacteria (chapter 13), because these often thrive in sulfury environments.

Sulfur caves are not very accessible, nor are deep sea sulfur springs and seeps. If you do happen to have an opportunity to view either (perhaps through connections with a professional researcher), the information in chapters 5 and 9 should be helpful.

GROCERY STORES, RESTAURANTS, AND KITCHENS

Be inspired by chapter 10 and then seek out grocery stores and restaurants where savory foods are being prepared. At the very least, eat interesting cheeses! Enjoy the many beverages enhanced by bacteria! Try making some foods and beverages yourself. Get specialized cookbooks or make contacts with people who are brewing, baking, or pickling and learn from them. In addition to all of the comestibles described in chapter 10, there are some foods and drinks enhanced by vinegar, courtesy of the alpha proteobacteria described in chapter 6. Also, some cuisines use cyanobacteria as food (chapter 14). If a food or sauce seems quite salty, consider the participation of salt-loving bacteria (chapter 4).

By avoiding bland or highly processed or chilled fresh foods, you may already be venturing into the realm of microbially enhanced cuisine. Suggestion: Have a dinner party based entirely on foods and drinks embellished by bacteria. Even if you are somewhat cautious, you can put together a nice meal (or appetizer) of aromatic cheeses, olives, sourdough bread, and lambic beer (or certain wines). A supplement to understanding and appreciating bacterial foods is to understand what happens to the food after we eat it. Consider reading about intestinal gases (in chapters 2 and 9) as well as about the bacteria of our digestive system (chapters 12 and 15). Many of our symbiotic bacteria are the same as or closely related to the bacteria that ferment foods—which is probably how they were first introduced to our food.

ANIMALS

In this guide to bacteria, other organisms (e.g., animals) are mentioned mostly as hosts for bacterial symbionts. Chapters 12 and 14 are devoted almost entirely to such symbioses. It is suggested that you work backward in planning a field trip to look at animal hosts; use other guide books to familiarize yourself with the animals that you expect to see, and also review these two chapters. In general, if an animal is eating plants (or algae), it has bacterial symbionts helping it to digest cellulose. The digestive systems of animals are wonderful bacterial habitats. To learn which bacteria are producing various intestinal gases (of carnivores, herbivores, or omnivores), see chapters 2 and 9. Also, nearly every moist or transiently moist surface of every animal is a habitat for bacteria. Therefore, although parts of chapter 12 seem to be specifically about humans, you could substitute nearly any other animal. For additional information on other symbionts, see chapters 15 and 16. Keep in mind also the gamma proteobacteria that are symbionts of shipworms and the bioluminescent gammas that are symbionts of many deep-sea organisms (chapter 8). Sulfureta abound in symbiotic associations as well (chapter 9), some of which may be depicted in museums in dioramas, for example of deep-sea vents.

Zoos and aquaria provide opportunities to observe the challenges of maintaining wild animals in captivity. Often, these challenges include keeping the bacteria happy. In zoos, some fastidious herbivores (e.g., three-toed sloths) are difficult or impossible to maintain because they and their symbionts are accustomed to specific diets and habitats. Most aquaria maintain a biological filter of nitrate-using bacteria (chapter 6) to keep the tanks clean and do regular battle against films of cyanobacteria. Also, herbivorous (actually algivorous) fish (and their symbiotic bacteria) present some challenges in feeding, often requiring supplements to their diets.

PLANTS AND FUNGI

To view bacteria–plant or bacteria–fungi (lichen) symbioses, you need to be fairly confident about your ability to identify plants and lichens. Plan to work backward—using guides to the plants and lichens of the habitat you are in, and looking for specific ones that are known to have symbionts (based on your reading of this field guide). Many plants contain nitrogen-fixing bacterial symbionts; see chapters 6 (on *Rhizobium*), 11 (actinomycetes), and 14 (cyanobacteria). Plant galls are a somewhat difficult subject for amateurs because it can be difficult to determine whether they were caused by bacteria, viruses, or fungi. Nevertheless, galls on herbaceous plants and trees are fascinating and worth seeking out. Also, look for cyanolichens and for plants growing at the edges of bogs that show the effects of metal toxicities due to bog bacteria (chapter 7). In the bog itself, you might find insectivorous plants, some of which are assisted in digestion by bacteria, or which may be digesting the bacteria themselves.

FIELD TRIPS TO SPECIFIC HABITATS

FARMS (INCLUDING COTTAGE INDUSTRIES)

Look for bacterial participation in pest control, fertilization (as in legumes), and the husbandry of domestic herbivores. Cud chewing and gaseous belching by ruminants, as well as the odors of chicken houses, compost heaps, and freshly turned soil, are all bacterial field marks. The making of silage and the careful avoidance of green hay in lofts both require a consideration (even if unconscious) of bacteria. Frost damage to plants may have bacterial aspects, along with some precipitation, such as snow! Farm work may include cottage-industry activities such as using traditional (bacterial) methods of preserving food and drink and traditional cloth-dying and linen-making techniques. In contrast, farming on an industrial scale might involve smelly waste lagoons (remediated by bacteria), eutrophication of water bodies (encouraging a diversity of bacteria), and soils that have become salty (encouraging halophiles) due to excess fertilizer. Special feedlot diets (and doses of antibiotics) may be detrimental to the bacterial symbionts of ruminants.

URBAN SETTINGS

Look for mosses, enhanced by the presence of cyanobacteria, on sidewalks and roofs. Any damp, shadowy wall or fountain is likely to have cyanobacteria. Look for bacterial "decay" of some stone monuments, including those in cemeteries. Ornamental plants in parks or in florist shops may be keyed out, and some (e.g., legumes and cycads) may be found to be symbiotic with bacteria. Sewers and Dumpsters may be sources of methanogens. Grocery stores, especially ethnic or gourmet ones, may have interesting cheeses and other fermented foods and drinks.

INDUSTRIAL AREAS

Watch for signs of methanogens, the producers of landfill methane and "sewer gas." Also, note any corrosion of concrete, stone work, and metal—a complex phenomenon that often has a bacterial component. Some industries are attempting bacterial remediation of wastes that go beyond the usual sewage treatment. These include clean-up of oil spills and reclamation of metals in mine waters. Some bacterial processes have been modified for the purpose of making beers, cheeses, and dyes on a commercial scale. Strict control of microbes or appropriation of microbial-type chemical reactions may often be seen in such processes. Furthermore, any genetic engineering of organisms using bacterial genes may be viewed as a direct appropriation of bacterial activities.

AQUATIC COMMUNITIES

All microbial ecology is essentially aquatic ecology. A field trip to any watery habitat is likely to involve the use of many chapters of this field guide. Take the following list of destinations, and use it along with the index to plan specific field trips to sources and bodies of water:

1. Fresh waters (lakes, ponds, streams, rivers)
2. Karstic bodies of water, with or without the presence of sulfides
3. Iron-rich waters (these are likely to show signs of manganese activity as well)

4. Marine waters (coral reefs, intertidal flats, estuaries, rocky and sandy shores)
5. Alkaline (soda) waters

Note that hot springs and hypersaline waters are mentioned in the first section of this chapter because they are the subject of entire chapters. Also, wetlands may be flooded to various degrees, some making the transition to "body of water" on a seasonal basis. Even the driest of habitats listed here, including deserts and dunes, may be viewed as microbial communities waiting in dormancy for moisture via rain or flood.

DUCKWEED

It happens that duckweed (a floating aquatic plant) is mentioned in three different places in this field guide, and it might be considered a very specific focus for a field trip as long as one has access to a microscope. Chapter 7 recommends the waters and sediments beneath duckweed as a source of magnetotactic bacteria. Chapter 16 suggests looking for tiny freshwater clams among duckweed and breaking them open to look for spirochetes. And chapter 18 describes communities of microbes attached to plants (including duckweed), rocks, logs, and the surface films of water. Although they are challenging to see even under a good microscope, some community members are likely to be stalked bacteria of the planctomycetes and other stalked microbes.

A bit of duckweed under the scope is likely to be a rewarding experience in general, and is good practice for microscopy even if you cannot see any bacteria. Often an abundance of ciliates and rotifers or other eukaryotic microbes may be observed. Also rewarding—even if it yields mostly eukaryotes—is some of the surface film of a still pond or lake.

WETLANDS

Wetlands of all kinds (such as bogs and swamps) are highly recommended as field trip destinations. Keep in mind the wide range of habitats to explore: iron-rich waters and sediments, aerobic and anaerobic habitats, and many interesting symbiotic associations, such as those of carnivorous bog plants.

CAVES, CLIFFS, AND DAMP ROCK LEDGES

Look for cyanobacteria on the rocks and just under the surfaces of rocks. For more exotic bacteria, use the index to locate information on troglodytic archaea and sulfur bacteria of unusual caves.

DESERTS AND DUNES

In either deserts or dunes, look for the cyanobacteria (and other microbes) of "desert crust." Whatever plants are surviving in the relatively poor soils are often highly dependent on nitrogen-fixing bacteria. Also, depending on which rocks and sediments are present, you might see endolithic cyanobacteria, halophilic bacteria, and desert varnish. Keep in mind that any body of water, including temporary ones such as the damp swales between sand dunes, is likely to have an active and complex microbial community.

TEMPERATE FORESTS

Forests, by definition, are dominated by trees and may be among the more difficult of habitats for microbial field trips. Try focusing on symbioses, such as those between nitrogen fixers and plants or between bacteria and herbivores. Forests are good places to view the effects of microbial decomposition (e.g., leaf litter, leaf mold). However, much of this decomposition may be dominated by the activities of fungi.

TROPICAL FORESTS

From a temperate point of view, everything in the tropics is more so—more niches, more competition, more biomass, and more species diversity. You are, therefore, forgiven if you are looking everywhere but at the bacteria, instead focusing on the dense tropical forests with layers and layers of large complex organisms. If you are investigating the ground, though, keep in mind that cycles of decomposition are quicker, resulting in thinner soils. Fungi rather than bacteria seem to be the major decomposers and major displayers of their activities. As with temperate forests, you might focus on symbioses between bacteria and plants (e.g., bromeliads) and between bacteria and animals (especially herbivores).

FIELD TRIPS THAT FOCUS ON BIOGEOCHEMICAL CYCLES

A biogeochemical cycle is the sequence of chemical reactions by which a particular element is converted from one form to another as it passes through organisms, bodies of water, the atmosphere, and various geological features. The usual representation of these cycles is a complicated set of loops showing many different possibilities. Some of the more important cycles are those of carbon, nitrogen, sulfur, phosphorus, and iron.

A complete plan for a field trip should include the use of an ecology textbook and its diagrams of specific cycles. Be assured that most parts of the major cycles are mediated in whole or in part by bacteria. Strategic use of an ecology text along with the index of this book can help you plan a far-flung and perhaps lengthy field trip to view as many parts as possible of one of the cycles. Many ecology books do not mention bacteria as extensively as they might, given their focus on biogeochemical cycles. Often bacteria and fungi are discussed together as "saprophytes" or "decomposers." However, even parts of cycles attributed (by some ecology texts) to plants and animals may be readily extrapolated to bacteria, including cyanobacteria and other autotrophs and the myriad fermenters that make herbivorous life possible.

FIELD TRIPS ORGANIZED AROUND COLLECTIONS

INSECT COLLECTIONS

Focus on insects with bacterial symbionts, looking in particular at the chapters on proteobacteria as well as in the index. Gall makers are an interesting focus for an insect collection, along with the galls themselves. Some are formed via bacterial interactions, although many others involve fungi or viruses.

FOSSIL COLLECTIONS

The index of this guide may help you arrange a fossil collection with a bacterial theme. Key words include microfossil, stromatolite, death mask, trilobite, slate, and limestone. Keep in mind that most of the history of life (4 billion years) has involved bacteria.

SHELL COLLECTIONS

Use the index to focus on those organisms with bacterial symbionts or with strategies for coping with sulfidic environments. These include *Mya*, *Ensis*, plumed worm cases made up of mollusc shells, lucinids, thyasids, and solemyids.

PLANT OR LICHEN COLLECTIONS

Focus on those organisms with agrobacterial or nitrogen-fixing symbionts. It may be possible to make dried, pressed, or boxed herbarium mounts of some bacterial structures of plants. A lichen collection might focus on those with cyanobacteria.

UNCONVENTIONAL OR IMAGINARY FIELD TRIPS

Expeditions to the North or South Poles, voyages in deep-sea research submarines, treks to remote places accessible only with guides, visits to deep mines and caves not open to the public, and other such adventures are not described here, even though such areas are full of bacteria. Obviously if you or I have an exceptional opportunity for an exotic field trip, as passionate amateur naturalists we will take full advantage of the opportunity by scouting guide books of all sorts. Should the opportunity arise, selective use of this guide would serve as a useful introduction. If remote parts of the planet are difficult to reach, however, anyone can explore other planets or fantasy landscapes—even if just in the imagination.

OTHER PLANETS

In seeking life on other planets, it is wise to consider bacteria rather than animal-like organisms that somehow resemble or behave like us. Most life is bacterial, and only bacteria have colonized the Earth with such diversity and versatility. Animals are an anomaly on our own planet; they are far outnumbered by microbes and are relative latecomers in the history of life on Earth. We humans have occupied the planet for a mere 50,000 generations (20 years per generation for about a million years) and, despite our growing population, we are still rather scarce in most environments. For example, we have an almost complete lack of presence in and on the ocean, which occupies three quarters of our planet.

Some Earthly bacteria are reasonable prototypes for the sorts of organisms that we might look or test for on other planets. Indeed, when planetary geologists consider which planets and moons might be capable of supporting life, they are often thinking of life below the surface of the planet, most likely on a microscopic scale. The Earth is unusual in its abundance of water and moderate temperature, a result of its distance from the sun. Most planets (and moons) are too cold (or hot) at the surface to have liquid water, but some may have molten interiors near which life might dwell. Subsurface conditions could exist in which heat-loving bacteria (some of them chemoautotrophs) might support a community of microbes (and possibly larger creatures) similar to those of deep-sea hot springs. Also, cold, even subfreezing, brines beneath some planet surfaces could potentially harbor slow-growing halophiles. Try reading chapters 1–4, 7, 9, and 17, keeping in mind that extraordinary conditions on Earth—boiling, acidic springs, for example—may have counterparts on other planets.

BACTERIA IN FAIRY TALES AND FANTASY LITERATURE
Wildly sprouting agrobacterial swellings of trees are often depicted in haunted woods, such as in the illustrations of Arthur Rackham. Odd smells and exudates (miasmas) of swamps and bogs are often bacterial, and marsh gas that sometimes bursts into mysterious flames is a product of methanogens. The very nature of bogs (with their eponymous bogeymen) lends itself well to fairy-tale settings. Surely the source of iron tools used by dwarves, for example, would have been bacterial bog iron. Some gigantic bacterial colonies such as those of *Nostoc* have fascinated humans, as evidenced by the numerous colloquial names for the colonies and the prevalence of slime in fantasy settings. Furthermore, the gigantic animals of both past and present are almost always herbivores, in debt to their digestive bacteria. Authors inventing gigantic animals might keep in mind that not all animals should be made ferocious and predatory! The circumstances of bacteria are so weird and otherworldly that this entire field guide could be read with fantasy or fairy-tale literature in mind.

CHAPTER 1

Ancient Hyperthermophiles and Thermophilic Green Nonsulfurs

It was hot when life originated on Earth about 4 billion years ago. Only half a billion years before this, the planet had begun as a sphere of molten rock. Gradually, temperatures dropped, a surface crust solidified, and rain began to fall from the cooling skies, filling shallow seas. Active volcanoes and thermal springs became abundant. Collisions with meteorites and fragments of asteroids and comets were also probably frequent. The atmosphere must have been filled with debris from these volcanoes and impacts, resulting in a rather hazy, reddish, or even dark sky, and possibly a greenhouse effect that maintained the warmth in the atmosphere. The first land masses may have been volcanoes rising up from the seas.

THE FIRST 500 MILLION YEARS OF LIFE

Life is hypothesized by some scientists to have originated and begun to evolve in boiling or near-boiling springs. No oxygen was present in the early atmosphere—which is fortunate, since the first chemical building blocks of life would never have held together in the presence of this highly reactive, destructive gas that without special processing is toxic to all organisms. Although humans, along with many other organisms, are breathing oxygen and would die without it, our relationship with that gas is a delicate one. Respiration seems to have evolved about $2\frac{1}{2}$ billion years ago from mechanisms that detoxified oxygen. Because oxygen respiration is also a marvelous way of assimilating energy, oxygen-respiring organisms have been quite successful in terms of both distribution and number of species. However, all oxygen-respiring, or aerobic, organisms require numerous other mechanisms to contain and control oxygen. Many good hypotheses that explain why we age and die revolve around the cumulative effects of oxygen damage to our DNA and cells. Certain diseases may be attributed to oxygen damage as well. Ultimately, oxygen is a poisonous gas. Properly used, it is a slow poison, and the benefits of oxygen respiration seem to outweigh oxygen damage.

One theory posits that the earliest organisms (bacteria or bacteria-like cells) were thermophiles (literally, heat lovers)—and extreme ones at that, thriving at boiling or near-boiling temperatures. They were of necessity strict anaerobes, requiring no oxygen and perhaps even lacking any natural defenses against oxygen. This theory fits well with classifications of modern bacteria, which suggest that present-day anaerobic hyperthermophiles (extreme heat lovers) have retained more characteristics of ancient bacteria than have other bacterial groups.

Exactly how life originated is a subject of great speculation. In brief, there seems to have been a transition stage in which small building blocks of life, such as amino acids and nucleotides, formed into larger molecules—proteins and RNA—with the help of energy sources such as the heat of thermal springs. This much can actually be accomplished in laboratroy experiments in which hot spring conditions are simulated. The next step by which proteins, RNAs, and other large molecules "self-assembled" into replicating, energy-using, enclosed systems (that is, *life!*) has not been accomplished in any laboratory, although various aspects of these processes have been replicated. There remains considerable room for experimentation before that remarkable transition from chemistry to biology is well understood and can be demonstrated in a laboratory.

It is clear, however, that life indeed originated, and it is interesting to hypothesize about what those first organisms were like. Their immediate environment may have been strangely abundant in food molecules—amino acids, nucleotides, proteins, and RNAs. Any molecules that had not somehow assembled into organisms would likely have been among the most directly available and accessible sources of energy and materials (food). It was an unusual period of Earth's history and one not likely to be repeated again; in today's environments, anything resembling a food molecule is quickly scavenged by hordes of microbes. Waiting around for food to appear in the environment is not a good modern strategy for nutrition, unless the waiting period is spent in a dormant form. However, when life was still a rarity and edible molecules were in some abundance, that might have been the simplest and most effective strategy. Therefore, the earliest life forms—the anaerobic hyperthermophiles—may also have been consumers of food molecules from their environments, that is, "heterotrophs." Certain types of autotrophy (the synthesis of food) may also have been important at this time.

HYPERTHERMOPHILES
THE MOST ANCIENT BACTERIA

Several groups of bacteria that thrive in hot water (over 45°C [113°F]) are called "thermophiles" (heat lovers). At temperatures greater than 60°C (140°F), eukaryotes do not grow. Therefore, any activity at temperatures over 60°C may be considered a good indicator of bacterial life, especially if scums, slimes, or flocs are found. Exactly which bacterial groups are present may be indicated by specific smells, colors, and textures and by a further refinement of temperature range. Between 60 and 80°C (140 to 176°F) may be found colorful mats of thermophilic green nonsulfurs (described in this chapter) and cyanobacteria (chapter 13). Bacteria that thrive at temperatures above 80°C (176°F) are called "hyperthermophiles." This group includes the ancient hyperthermophiles (genera *Thermatoga*, *Hydrogenobacter*, and *Aquifex*; this chapter), the hyperthermophilic archaea (chapter 3), and species in the genus *Thermus* (chapter 17).

This chapter focuses on two groups that seem to have taken fewer detours during their evolutionary path from the first bacteria and to have retained more of their ancestral genes than other groups: the ancient hyperthermophiles and the thermophilic green nonsulfurs. Both live in the types of extreme environments that were typical of the early Earth, and to detect their field marks, it is necessary to journey to environments considered intemperate by humans. Ordinary warm springs will not do; the temperatures and conditions must be extreme.

WHERE TO LOOK FOR ANCIENT HYPERTHERMOPHILES
Many of the ancient hyperthermophiles tend to live in marine hot springs that are usually much too deep to be accessed except by

research submarines. However, you can visit a hot spring environment on land, and as long as you have chosen one hot enough (over 80°C [176°F]), you can imagine that you are in the vicinity of something like the most ancient microbial communities. Yellowstone National Park, in Wyoming, is the best place to find this sort of environment. Three other major thermal areas, in Iceland, New Zealand, and eastern Russia, have been altered to gain thermal power or are less well exposed and therefore exhibit a less diverse display of heat-loving microbes. Other locations for hot springs of this temperature in the United States include sites in California, Utah, Oregon, Colorado, New Mexico, and Idaho. Hot springs (actually steam vents) greater than 80°C (176°F) may also be found in Washington and Hawaii, although these are often inaccessible because they are associated with active volcanoes. At hot springs cooler than 80°C, the microbial community may include bacteria of more recent lineages; their colors will tend to obscure your "view" of the ancient hyperthermophiles.

Yellowstone National Park

Yellowstone National Park should be considered a primary destination for any field microbiologist, amateur or professional. The park encloses and protects the largest pristine assemblage of hot springs, geysers, fumaroles (steam vents), and boiling mud pots in the world (table 1.1). A visit to Yellowstone is the best way to view a diversity

TABLE 1.1. Where to Look for Field Marks of Hyperthermophiles in Yellowstone Park, Wyoming

---

Lower Geyser Basin
  Octopus Spring (84–91°C [184–196°F]). This is a well-studied research area (see chap. 17).
Midway Geyser Basin
  Grand Prismatic Spring (64–87°C [147–188°F])
Black Sand Geyser Basin

Source: Carl Schreier, *A Field Guide to Yellowstone's Geysers, Hot Springs, and Fumaroles* (Moose, Wyo.: Homestead, 1992).
Note: These features are likely to have a range of temperatures, and it may not be possible to get close enough to take samples. At Yellowstone, water boils at 93°C (199°F) because the park is 7,500 feet above sea level.

of bacteria in thermal and hyperthermal environments. The U.S. Park Service has built miles of boardwalks so that many of the thousands of thermal features may be viewed up close safely and without disturbing the delicate balance of microbiology and geology.

Most of the area of the park consists of the collapsed center of a huge volcano, or caldera, that first erupted about 2 million years ago, again 1.2 million years ago, and most recently 600,000 years ago. It is still considered active and will erupt again. The myriad thermal features in and around the caldera are testament to the ongoing turmoil just beneath the thin surface crust.

Yellowstone Park is one of the few places in the world where one may look out at a vast panorama of bubbling, hissing, erupting, oozing fluids and minerals. This scene is probably as close as we can come to a view of what the Earth's surface was like 4 billion years ago. Such a view is possible from the Visitors' Center overlooking the Porcelain Basin in Norris Geyser Basin, one of the hottest areas of the park (plate 1).

Planetary geologists have hypothesized that the surface crust of the early Earth was much thinner than it is now, and therefore more active with earthquakes and thermal features of all kinds, including volcanoes. The crust at Yellowstone may be only about 40 miles thick rather than the more typical 90 miles for the rest of the Earth. Furthermore, Yellowstone sits atop a slowly moving hot spot (inching northeast) from which magma deep within the mantle has extended upward and has broken through parts of the Earth's crust. As a result, molten rock may be just 2 to 3 miles below the surface in some places. The whole park may be considered a rare window on the geothermal world below the Earth's surface, and as such is an equally rare window on what the Earth might have looked like when life originated.

FIELD MARKS AND HABITATS
OF ANCIENT HYPERTHERMOPHILES

In general, temperatures above 80°C (176°F) and neutral to alkaline conditions are reliable field marks for the few bacterial genera considered to be ancient hyperthermophiles: *Thermatoga*, *Aquifex*, and *Hydrogenobacter*. (See chapter 3 for another ancient branch, the hyperthermophilic archaea, which are found in *acidic* hot springs, and chapter 17 for *Thermus*, another heat lover.) All other organisms are excluded at these extreme temperatures. Any slimes and some of the colors such as pinks and yellows that you observe above 80°C are fairly reliable bacterial field marks. Concentrate on hot springs that are stable enough—neither eruptive nor churning—to allow visible bacterial assemblages to form. Be aware, however, that colorful sediments at temperatures over 80°C may also be due to mineral precipitates such as red iron and yellow sulfur. The following are good rules of thumb for distinguishing minerals from bacteria:

1. Some minerals grow in from the edge of a feature, while bacteria spread out from the source of the heat.
2. Minerals tend to be crusty or crystalline and rigid, whereas bacterial assemblages may appear to shimmer or "wave in the breeze" in flowing water. In general, bacterial growths are softer than mineral deposits.

Determining the temperature of a hot spring likely to have hyperthermophiles is the next challenge in identification. Usually, you should not get close enough to a boiling spring or geyser to be able to insert a thermometer. The surrounding ground may be thin and unstable, and a fall into boiling water may be lethal. Instead, take advantage of any signage indicating temperatures, as well as the knowledge of the park rangers. Carl Schreier's *A Field Guide to Yellowstone's Geysers, Hot Springs, and Fumaroles* also provides temperature information, although this is highly subject to change according to the season and because of the instability of most features. For taking temperatures at a distance, consider an infrared temperature sensing "gun." If you use a regular thermometer, do not use one containing mercury, as breakage will release this dangerous pollutant. Make sure that the thermometer registers boiling temperatures.

One of the most interesting ways of determining temperature is by "colorimetrics"—that is, using certain-colored bacteria or a lack of them as indicators of particular temperatures (table 1.2). Below 60°C (140°F) eukaryotic organisms may be found. For example, *Cyanidium*, a bright grass-green alga, lives at temperatures

TABLE 1.2. Colorimetric Determination of Temperatures in Neutral to Alkaline Hot Springs

Approximate temperature range	Color of water or sediment	Confounding factors
93°C (199°F)*	Bright-blue steaming water, often with dissolved silica and steam or vapor. Also white sediment in steamy water, due to silica minerals such as geyserite or sinter. (In Mammoth Terraces area, the white is travertine, or calcium carbonate.)	Bubbling may or may not indicate boiling; sometimes it is just escaping gas.
75°C (167°F)	Gradient of yellow-orange-brown-green (going from hot to warm).	In iron-rich or sulfur-rich areas, the colors could be due to minerals rather than bacteria; see text for hints on distinguishing the source.

* At the elevation of Yellowstone, this is boiling.

up to 55°C (130°F) in *acidic* waters, along with diatoms, a type of protist.

An especially effective use of table 1.2 is to locate boiling features (e.g., erupting geysers) in which a gradient of concentric colors may be observed in pools or in ribbons of color where hot water is flowing away from the feature (plates 2–4).

The focus for detection of ancient hyperthermophiles should be any white area of sediment flanked by boiling or near-boiling water on one side and colorful yellow-orange bacteria on the other. The thermal feature should be alkaline to neutral in pH. Some pink-colored filaments in such areas have been determined to be a species or relative of *Aquifex*. Be aware that in the same areas you may also find pinkish-yellow filaments, which are likely field marks of *Thermus* bacteria (see chapter 17). An acidic hot spring is likely to have a different assemblage of field marks and is more appropriate for detecting hyperthermophilic archaea (see chapter 3; full interpretation of a hot spring environment should include information from chapters 1, 3, and 17).

The ancient hyperthermophiles, as well as hyperthermophilic archaea and *Thermus* species, may be found at temperatures lower than the narrow range recommended for observing field marks. At these lower temperatures, they may be present and even thriving but obscured by other bacteria, especially brightly pigmented, photosynthetic ones.

VIEWING ANCIENT HYPERTHERMOPHILES UNDER A MICROSCOPE

In most cases it will not be possible or advisable for you to get close enough to the boiling or near-boiling waters of thermal springs to collect a sample. One exception might be tiny, unmarked features away from the usual tourist areas. For example, on a roadside near West Thumb Geyser Basin, small flows of hot water can be seen bubbling up on the shore of Lake Yellowstone (fig. 1.1). In such an area, you could determine the temperature and collect a small drop of water or speck of sediment for viewing under a microscope (fig. 1.2). To preserve your sample for later viewing, add formalin to make a total concentration of about 3% (3 parts formalin to 97 parts sample water). Many ancient hyperthermophiles are coccoids, rods, or filaments, and they are often large. Keep in mind that at lower temperatures, eukaryotes such as green algae and diatoms will obscure (or distract you from) the bacteria. Very small thermal features will probably not maintain a large temperature gradient, so your sampling area may be just a few millimeters.

If you have found a tiny, unmarked thermal feature and have about a week to spend in the area, consider submerging a glass microscope slide in the hot water or moist sediment and leaving it for a week. (Realize, of course, that the U.S. Park Service could

Figure 1.1.   The author and her daughter at a tiny thermal feature on the shore of Lake Yellowstone near West Thumb. A sample (the size of a speck) is being prepared for the field microscope.

rightly interpret this as the leaving of trash; consider asking permission or being very discreet, taking care to remove your experimental slide when done.) Microbiologists have had good success getting hyperthermophiles to actually grow on the glass, which may be wiped clean on one side and viewed beneath a coverslip on the other.

CULTURING ANCIENT HYPERTHERMOPHILES

Culturing hyperthermophiles involves maintaining boiling or near-boiling conditions in a complex medium, devoid of oxygen. This is not a simple amateur activity.

Thermophilic Green Nonsulfurs
A Fork in the Evolutionary Tree

Green nonsulfur bacteria are descendants of the first photosynthetic bacteria. Their branch on the family tree appeared soon after that of

FIGURE 1.2. Taking the temperature of the feature described in figure 1.1. The water bubbling up was boiling, and by standing too long right next to it, I let the soles of my sneakers melt a little! The sampled (orange) area shown here was at 65°C (149°F), a good habitat for green nonsulfurs. A little farther away, at 55°C (131°F), a green sample was full of eukaryotes such as diatoms.

the ancient hyperthermophiles. Green nonsulfurs love hot water too, although they are not true extremophiles (lovers of extreme conditions). Look for them in hot springs between 60 and 80°C (140–176°F). While some grow at cooler temperatures, so too do many eukaryotes that obscure your view of the bacteria. To make a reliable identification of green nonsulfurs, look for hot springs in the right temperature range (table 1.2).

The somewhat cumbersome name *green nonsulfurs* is derived from two major aspects of their metabolism. These bacteria are photosynthesizers that use a version of the green pigment chlorophyll to capture light energy. Unlike other groups of photosynthetic bacteria that appeared early on the tree of life, green nonsulfurs do not require sulfur compounds such as hydrogen sulfide (H_2S) as a source of hydrogen in their metabolism, nor do they use water (H_2O) as a source of hydrogen, as more advanced photosynthesizers such as the cyanobacteria do. Instead they use organic molecules.

THE EARLIEST PHOTOSYNTHESIS

Chloroflexus is the green nonsulfur most commonly observed in hot spring environments. These bacteria are used here as an example of some of the peculiarities of ancient bacterial photosynthesis.

All photosynthesizers—whether trees, eukaryotic green algae (pond scum), cyanobacteria, or green nonsulfurs—capture light energy using some version of the green pigment chlorophyll. They use that light energy to make food molecules (often sugars) from decidedly non-foodlike molecules, most commonly carbon dioxide (CO_2). Some form of hydrogen is also needed because food molecules invariably turn out to be some type of *hydro*carbon. For example, trees take water (H_2O) as a source of hydrogen to make the sugar glucose ($C_6H_{12}O_6$) from carbon dioxide:

$$CO_2 + H_2O + light \rightarrow food\ molecules$$

Most variations on photosynthesis center around those starting compounds. Water is a convenient, nontoxic source of hydrogen that is used along with carbon dioxide by all eukaryotic photosynthesizers and most cyanobacteria. The other photosynthetic bacteria as well as some cyanobacteria use other sources of hydrogen with carbon dioxide. Hydrogen sulfide is a favorite, although users of this compound are limited to sulfur-rich environments, which are fairly rare on present-day Earth.

The starting compounds used by *Chloroflexus* for photosynthesis make it seem rather strange, even by comparison with other obscure bacterial photosynthesizers. For *Chloroflexus*, the optimal source of carbon and hydrogen (including free hydrogen) is some small, available food molecule such as acetate or pyruvate. These molecules are presumably waste products of other bacteria, such as certain anaerobic heterotrophs that only partially digest their food. If oxygen is present, *Chloroflexus* acts as a heterotroph and consumes whatever small food molecules are available, through an unusual type of respiration. However, if oxygen is not present (as was the case on the early Earth), *Chloroflexus* can use light energy to make larger food molecules from smaller ones.

CHLOROFLEXUS AND THE EVOLUTION OF METABOLIC PATHWAYS

Evolutionists studying the early photosynthesizers are reluctant to form conclusions about the exact order in which different types of metabolism evolved. After all, we do not have the original organisms in hand. We can only look at their descendants, which have undergone almost 4 billion years of evolution. Surely there have been some changes in that time period! Nevertheless, the metabolic versatility of *Chloroflexus* and its position on the bacterial family tree—right before the divergence of most of the other bacteria—give clues to how bacterial evolution might have occurred.

Chloroflexus needs quite an accumulation of genes to accomplish its diverse metabolism, which includes photosynthesis under anaerobic conditions and respiration (heterotrophy) under aerobic conditions. According to one hypothesis, heterotrophy evolved before photosynthesis, and the ancestor of *Chloroflexus* was a fermenter (consumer of food). Metabolic processes tend to be complex, coded for by many genes. While individual genes might be lost or modified by mutation, it is difficult to lose an entire set of genes that codes for a complex function. Therefore, if the ancestor of *Chloroflexus* was a fermenter, then *Chloroflexus* might be expected to still contain genes for fermentation. According to this scenario, it is as if during evolution, *Chloroflexus* kept sets of old genes (like old furniture) in its attic or basement, rather than getting rid of them. However, this collection was not idle. Sometimes items were restored to use, combined in some new way, or even modified through mutation. This is how two new metabolic processes—photosynthesis and aerobic respiration—might have evolved in *Chloroflexus*. Indeed, some aspects of photosynthesis in *Chloroflexus* look like parts of fermentation run backward. By running such a reaction backward and applying lots of free energy from the sun, an anaerobic heterotroph could become an autotroph. (Yes, biological reactions can be run backward, with interesting consequences, as long as there is enough energy to force the reaction into reverse.) Similarly, aerobic respiration can be viewed as an elaboration of fermentation. Now imagine the ancestral *Chloroflexus* with bulging attic and basement maintaining a diversity of metabolic pathways, perched on the evolutionary tree just before the divergence of most bacterial groups. With various modifications and combinations of the *Chloroflexus* genes, almost all subsequent types of bacterial metabolism (featured in chapters 5 to 18) could have evolved. The peculiar heterotrophy of present-day *Chloroflexus* may have evolved following this divergence, after significant oxygen had become available in the atmosphere.

CHLOROFLEXUS IN THE FOSSIL RECORD

The most ancient microfossils and stromatolites, which are fossilized microbial communities dating from 3 to 3½ billion years ago, are usually interpreted to be cyanobacteria (see chapter 13). An alternative explanation, however, is that these are fossilized bacteria of the *Chloroflexus* type. This is a reasonable idea, given the fact that *Chloroflexus* bacteria seem to predate the cyanobacteria, but firm evidence as to the exact identity of early fossilized bacteria is not available.

FIELD MARKS AND HABITATS OF
THERMOPHILIC GREEN NONSULFURS

The cells of green nonsulfurs are long, thin filaments capable of a gliding type of motility. They are most visible when forming colorful gelatinous mats in alkaline to neutral hot springs between 60 and 80°C (140–176°F) (table 1.3). The mats range in color from yellow or beige to orange or orange-red as a result of carotenoid accessory pigments; in some cases they are green due to the presence of chlorophyll. Green nonsulfurs are fairly easy to see as long as they are any color but green (plate 5). Green mats of green nonsulfurs are difficult to distinguish from greenish or blue-green mats of cyanobacteria, which are often present in the same environment. Generally a mat of yellow- to beige-colored green nonsulfurs will be positioned below a thin, green to blue-green mat of cyanobacteria. If the hot spring is quite sulfury smelling, the mats may be inverted such that the cyanobacterial layer lies beneath and in more direct contact with sulfur compounds.

Be cautious in making macroscopic identifications. Sometimes cyanobacterial mats may become bleached to an orange color,

TABLE 1.3. Where to Look for Field Marks of Green Nonsulfurs in Yellowstone Park, Wyoming

---

Mammoth Hot Spring Terrace
    Opal Terrace, 160°F (71°C)
    Minerva Spring, 161°F (72°C)
    Orange Spring Mound, 157°F (69°C)
    New Highland Spring, 160°F (71°C)
    Canary Spring, 160°F (71°C)
Midway Geyser Basin
    Grand Prismatic Spring, 147–188°F (64–87°C)
    Turquoise Pool, 142–160°F (61–71°C)
Black Sand Geyser Basin
    Opalescent Pool, 144°F (62°C)
    Emerald Pool, 154.6°F (68°C)
    Rainbow Pool, 161°F (72°C)
Upper Geyser Basin
    Morning Glory Pool, 171.6°F (77°C)
    Beauty Pool, 164–175°F (73–79°C)
West Thumb Basin
    Abyss Pool, 172°F (78°C)
    Blue Funnel Spring, 172–182°F (78–83°C)

Source: Carl Schreier, *A Field Guide to Yellowstone's Geysers, Hot Springs, and Fumaroles* (Moose, Wyo.: Homestead, 1992). *Note:* Look for green-blue to green cyanobacterial mats with tan-orange bacteria directly above or below green nonsulfurs. Temperatures may vary, with edges and runoff areas likely to be cooler. You may not be able to take samples or insert a thermometer. Also check cooler runoff areas of features listed in table 1.1.

revealing their orange carotenoid accessory pigments. Green nonsulfurs can also form greenish or orange mats independent of visible cyanobacterial mats. The presence of obvious cyanobacteria (green to blue-green mats) with underlying or overlying gelatinous beige-orange mats in a temperature hot enough (60–80°C [140–176°F]) to omit eukaryotes may be the best combination of field marks for the green nonsulfurs (plate 6). Green nonsulfurs have also been observed in low-temperature marine and hypersaline microbial mat communities, although they are likely to be present in lower numbers and are difficult to distinguish from cyanobacteria.

VIEWING THERMOPHILIC GREEN NONSULFURS UNDER A MICROSCOPE

Do not touch or collect any colorful mat samples in any areas at or near the boardwalks in Yellowstone Park. This is not only dangerous but it is also a form of vandalism that can leave semipermanent marks. Rather, seek out tiny transient features (e.g., as shown in fig. 1.1) and take only a speck of sample, which is all you need for microscopy. If you wish to preserve the speck, place it in 3% formalin (3 parts formalin to 97 parts sample water). Make sure that the sample is from the correct layer in a neutral to alkaline thermal spring (60–80°C [140–176°F]). Compare samples from different layers to be sure. Look for long filaments capable of a gliding type of motility. The filaments may or may not be green, depending on the color of the layer you sample.

CULTURING GREEN NONSULFURS

Maintaining the right temperature and other conditions to grow cultures of predominantly green nonsulfurs is not a simple amateur activity. Green nonsulfurs may grow at more temperate conditions in a Winogradsky column (appendix A). However, it will be difficult to distinguish them from the other photosynthesizers that will proliferate.

SUMMARY: FIELD MARKS AND HABITATS OF ANCIENT HYPERTHERMOPHILES AND THERMOPHILIC GREEN NONSULFURS

Ancient Hyperthermophiles
- thermal features with temperatures greater than 80°C (176°F), alkaline to neutral pH (indicated by white geyserite [sinters]), and stable enough conditions to allow large assemblages of bacteria to develop
- soft, colorful assemblages (yellows-reds-pinks) that appear to flow or shimmer, radiating from a heat source (minerals, which may have similar colors, are harder and crustier and are usually deposited from the edge of a spring)

- long, hairlike strands, possibly pigmented (with pinks and yellows but not greens)
- white sediment flanked by near-boiling water on one side and yellow-orange bacteria on the other; cooler runoff waters may have orange or blue-green cyanobacteria

Thermophilic Green Nonsulfurs
- thermal features with temperatures of 60–80°C (140–176°F) (at cooler temperatures, green nonsulfurs may be confused with other organisms) and alkaline to neutral pH
- beige to orange gelatinous mats (sometimes greenish, but then difficult to distinguish from other photosynthesizers)
- presence of obvious cyanobacterial mats (green to green-blue) either just above or just below (if the cyanobacteria are orange or bleached, the green nonsulfurs will be difficult to distinguish from them)

Introduction to the Archaea
Methanogens, Hyperthermophiles, and Halophiles

The archaea certainly look like other bacteria, which is to say that they don't look like much of anything—tiny rods, spheres, and squiggles, with irregular, amorphous shapes perhaps a little more common. And indeed, the archaea, or archaebacteria, used to be grouped with the other bacteria, or eubacteria; in the 1970s, however, taxonomists elevated their phylogenetic status, classifying them in their own "domain," or "superkingdom," separate from other bacteria. (In this guide I use the term *bacteria* for both the archaebacteria and the eubacteria; some biologists, however, reserve the term for the latter group.) In addition to being promoted to a higher taxonomic level, the archaea have also become the subjects of much research activity. Until recently, this group was relegated to the fringes of bacterial study, perhaps because they are not pathogenic to humans and are found primarily in extreme environments, well out of the way of most human investigators. In fact the extreme nature of many species of archaea makes them candidates for participation in certain industrial processes that occur at high temperatures, pressures, acidities, or alkalinities. Many of the archaeal genes are being patented by investigators hoping to use them in various industrial applications.

It can be difficult to culture the extremophilic archaea, and it is even more difficult to imagine that they are living in some of the inhospitable environments in which they are found, such as boiling acidic water or crystals of salt. Distinctive field marks such as the generation of methane gas and pink coloration on salt crystals make archaea readily identifiable by field microbiologists, but for a long time they were not identified as a separate group. This changed (although it was accepted with reluctance by many microbiologists) in the 1970s when Carl Woese used DNA sequences to sort out parts of the bacterial family tree. Based on changes in DNA sequences that have been occurring slowly for the last 4 billion years, Woese was able to determine that the archaea comprise their own major branch on the tree of life (see fig. I.1). Analyses of entire genomes have revealed that the archaea share major gene groups with the other bacteria and with eukaryotes, though some genes are unique to the archaea.

The archaea are now known to constitute a major branch of organisms that diverged from eubacteria more than 3 billion years ago. Other characteristics of archaea now fit into the picture. It was known that their cell walls were strange, either fragmented or "missing" entirely and made of chemicals that are not typical of other bacteria. Furthermore, archaean flagella (their organelles of motility) are entirely different chemically from eubacterial flagella. This observation now makes sense in light of archaea's long past departure from the rest of the bacterial lineage.

Almost all known species of archaea are found in odd, inhospitable (from a human point of view) environments, and that characteristic is what makes these organisms so visible. Archaea are much more widespread than just extreme environments, but it is in the extremes that they can be seen unobscured by the more numerous eukaryotes or other bacteria. Exceptions? In some salt environments, the eukaryotic pink alga *Dunaliella* can thrive over populations of archaea; and the guts of many herbivores are teeming with eukaryotic symbionts as well as archaea and other bacteria. In most extreme environments, however, the archaea are free to bloom and display their field marks. If you are willing to venture into the extremes yourself, you will find them.

The following three chapters have been divided according to a scheme that uses ecological as well as phylogenetic characteristics. The methanogens (generators of methane; chapter 2) span more than one major branch of the archaeal tree, while the hyperthermophiles (heat lovers; chapter 3) and the halophiles (salt lovers; chapter 4) each occupy their own branch. This scheme does not do justice to the full spectrum of archaea, which includes deep subsurface and cold-loving species as well as many temperate species. However, it does fit with the emphasis of this field guide—bacteria with *macro*scopic field marks. An additional ecological grouping of archaea, the troglodytes, or cave dwellers, which dwell in deep cracks throughout the Earth's crust, is discussed briefly in chapter 3. Although they are abundant and diverse, troglodytes are not featured in their own chapter because of their inaccessibility to most amateur naturalists.

CHAPTER 2

Methanogens

Methanogens are archaea that produce the flammable, odorless gas methane (CH_4) as a product of their metabolism. Methane, even in the trace amounts present in our atmosphere, is considered to be a greenhouse gas. Some of it is geochemical in origin, and some is a waste product of human industry. However, the vast majority is generated by bacteria, in particular by the methanogens. Bubbles or emissions of methane are the primary field mark for this group and can be detected rather easily in many anoxic (oxygen-free) environments.

METABOLISM IN METHANOGENS

What methanogens are doing for metabolism is really rather strange, especially from a human-centered point of view. Some methanogens are autotrophs or have autotrophy as one of their pathways. This means that they are somehow getting sufficient energy and carbon to make the carbon compounds they need without ingesting complex food molecules. However, these are not typical autotrophs. Most autotrophs are photoautotrophs or photosynthesizers, using light as a source of energy and a fairly conventional pathway for making food. The autotrophic methanogens, however, are chemoautotrophs, using nothing more than carbon dioxide and hydrogen—gases that most organisms consider to be waste products. They channel some of the carbon from the carbon dioxide into energizing a large building-block molecule, acetyl coenzyme A, which is useful for making many other compounds. The rest of the carbon dioxide combines with hydrogen gas and eventually produces the waste product CH_4 (methane gas) as well as an energy-rich molecule, ATP (adenosine triphosphate).

These simplified equations compare photoautotrophy (as carried out in plants, algae, and cyanobacteria) with chemoautotrophy (as carried out in some methanogens).

Photoautotrophy

light energy + CO_2 + H_2O → sugar + O_2 (waste)

Chemoautotrophy

chemical bond energy + CO_2 + H_2 → sugar + CH_4 (waste)

Some methanogens are either heterotrophs or use heterotrophy as an alternative to autotrophy if the right food (such as acetate or some other simple carbon compound) is available. These food compounds are ultimately broken down to methane, with energy being transferred to the organism. Either way—whether by autotrophy using carbon dioxide and hydrogen gas, or by heterotrophy using acetate as food—it is a difficult life. Hydrogen gas is a scarce commodity in our highly oxidized world. Whatever hydrogen comes in contact with oxygen or any oxidized compound quickly becomes oxidized, often yielding water (H_2O). Furthermore, other organisms compete for acetate and hydrogen gas. Methanogens do best in environments that are devoid of oxygen and are home to heterotrophic fermenting bacteria that produce hydrogen gas, carbon dioxide, and acetate as waste products. Indeed, these other organisms are so essential to methanogens that the latter can be considered to be in obligate symbiotic relationships with some fermenters. One famous symbiotic relationship is described by this sequence:

		acetate		
		or		
carbohydrates	→	hydrogen	→	methane
(such as dead	via	or	via	
plants)	fermenting	carbon	methanogens	
	bacteria	dioxide		

An organism that was originally named *Methanobacillus omelianski*, for example, was demonstrated to be in fact *two* bacteria: a fermenter that uses ethanol as food and produces acetate and hydrogen gas as waste, and a methanogen that quickly uses those wastes and produces methane. In general, if there are methanogens present, the surrounding environment must be anoxic (even if only in the immediate locale), and there must be some source of hydrogen gas. If hydrogen gas is not available as the waste product of fermenters, it must arrive from some geological process such as a spring carrying anoxic waters from deep within the Earth.

Thus, the environments in which methanogens are likely to be found, and their field marks detected, include stagnant, anoxic waters and sediments in which plant material is being fermented (decayed); the anoxic parts of the digestive systems of organisms, especially herbivores; and springs, especially geothermal ones, that bear anoxic waters. Some trees and other plants may harbor methanogens; sewage, especially in home septic systems and some landfills, may also have distinctive methanogen field marks.

FIELD MARKS AND HABITATS OF METHANOGENS

ANOXIC SEDIMENTS

Find some nice stagnant, swampy water, and stir up some of the sediments with a stick. If you see bubbles of gas rising to the surface, that gas is most likely methane, the product of methanogenic archaea (plate 7). This gas should not smell sulfury. Although methanogens are ubiquitous in all anoxic habitats, whether freshwater or marine, at a wide range of temperatures and salinities, the most unequivocal observations of methane can be made in freshwater environments. In other environments, other gas-generating bacteria such as hydrogen sulfide–producing species may be dominant. Also, be sure that you are not stirring up a layer of cyanobacteria (visible as blue-green scum) or eukaryotic algae (green scum), both of which produce oxygen bubbles.

If you want to be absolutely sure that the bubbles are methane, choose a very stagnant site with a thick bottom of sediments, perhaps already bubbling a bit even without the aid of your stick. Then try Alessandro Volta's experiment, which he did in 1776 at Lake Maggiore: hold a candle (or a lighter) close as you stir up deep sediment (or plunge a length of pipe into the sediment to release trapped gases). Look for the flame to flare up a bit, indicating what Volta called "combustible air." Volta essentially discovered a new gas, methane, as well as what was later recognized as a field mark of an organism.

One modification to Volta's experiment includes plugging the end of a funnel, filling it with water, and holding it wide end down on the surface of some swampy sediments. You then stir the sediments around and beneath the funnel with a stick, or trample on the sediments with your feet to release methane, which will bubble into the funnel and displace the water. The methane may be released and lit by unplugging the funnel (plate 8). (Those who grew up in less litigious times may also recall, along with playing with BB-guns, carving with knives, and playing unsupervised in woods and ponds, capturing "marsh gas"—methane—in jars and lighting it for fun.) Note that oxygen is flammable too, but since we already have lots of it (20%) in our atmosphere, whatever flaring up you observe can be attributed to methane.

In some cases, swamp methane (or "swamp gas") lights spontaneously, forming pale, softly glowing balls of flame. These have been observed in many boggy and swampy areas and given colloquial names such as "will o'wisps," "ghost lights," "corpse lights," and "jack-o-lanterns." The phenomenon was first reported by Pliny the Elder in *Natural History* and called "inflammable mud," or *limum flagrantem*.

Enormous amounts of methane are produced constantly in swamps, bogs, tundras, marine sediments, and other anoxic environments. These environments are a primary source of methane in our atmosphere, although only trace amounts of the gas that is released can be detected. The rest is oxidized either by methanotrophs (see chapter 6) or by our extremely oxygen-rich atmosphere, becoming carbon dioxide. An interesting form of oxidized methane is the methane hydrate found in some deep ocean sediments. This is methane produced by methanogens and then "trapped" under high pressure and low temperature in a sort of "matrix" of water molecules. Thus, methane hydrate is a field mark of some deep-sea methanogens but is observable only by means of research submarines. It is also an indicator of a newly discovered deep-sea community of organisms that consists of not only bacteria but also pink worms and probably other creatures that feed on the bacteria.

AQUATIC PLANTS

Water lilies (*Nymphaea*) and several other aquatic plants, such as spike rush (*Eleocharis*), pickerel weed (*Pontederia*), arrowhead

(*Sagittaria*), and cattail (*Typha*), transfer large quantities of methane from their roots up through their leaves and stems (plate 9). At least in part, this methane transfer is enhanced by the airy spaces that characterize the tissues of these plants, facilitating both flotation and the transfer of oxygen and carbon dioxide. Cut open a cross-section of a leaf or stem, and you will see the lacework of open tubes.

The bubbles emitted by these water plants are field marks of methanogens—identifiers that in this case can be both seen and heard. As cattails or water lilies warm up in the morning, gas pressure increases, sometimes causing a crackling or popping sound as the drier leaves rupture from the gas pressure and bubbles. The plants may be shaken to encourage the release of bubbles. A popping sound may also be detected at night, as methanogens are better able to produce methane without the interference of oxygen production by daytime photosynthesis. Although methane is largely produced at night, most of it is not released via plants into the atmosphere until daytime, when it flows out along with oxygen and other gases produced during photosynthesis.

Cultivated rice paddies (*Oryza*) are really in their own category of methane emission, since rice is such an important crop and is grown in vast areas of water-logged soil, ideal for the production of methane. Along with the digestive tracts of herbivores, stagnant bodies of fresh water (swamps, bogs, marshes), and the burning of biomass, the sediments and waters of rice paddies are one of the top sources of methane in our atmosphere via their methanogenic inhabitants.

WET WOOD

Some trees, especially those that grow in water-logged soils and have water-soaked roots, trunks, or branches, exhibit the symptoms of "wet wood" (or "slime flux"), which is another field mark of methanogenic activity (fig. 2.1). The trees most susceptible to this condition are poplar, cottonwood, and elm. (Indeed, all poplars and cottonwoods and most elms may have this condition to some degree.) Less-affected species include birch, willow, maple, apple, and mountain ash. Wet wood is characterized by a colorless liquid or slime that seeps from cracks and wounds in the bark, darkens on contact with the air, and smells rancid or sour and forms bubbles. These symptoms, which are most visible in mid-summer, are caused

FIGURE 2.1. "Wet wood," caused by nonpathogenic methanogens and fermenters of the interior wood of some trees, is characterized by a damp leakage through fissures in the bark. There may be a somewhat rancid smell, which is caused by the waste products of the fermenters. Methane is also being released, although this is difficult to detect unless internal pressure produces enough gas to be lit.

by methanogens feeding on carbohydrates; they are usually not life-threatening to the tree.

Wood that is infected by these methanogens does not actually rot and, because it is home to methane-producing bacteria, may in fact be protected from rot by other organisms. Products of the methanogens are fatty acids (which become rancid), carbon dioxide, and methane—the same compounds produced by methanogens in the digestive systems of herbivores. The production of gas in wet wood is testament to how resourceful the methanogens are in exploiting sources of carbohydrates. In some cases, the pressure of methane gas being emitted from a tree wound (or from a hole bored by a forester) is great enough that the methane can be lit by a match, forming a jet of flaming blue gas.

DIGESTIVE TRACTS OF HERBIVORES

A major source of methane is the digestive systems of animals. It is not easy to calculate how much methane is emitted into the air from the mouths and anuses of ruminants, termites, and other animals, but it is estimated to be a major contribution to the atmosphere. Among the herbivorous and especially wood-eating or soil-eating arthropods, termites are the most reliable indicators of methanogens. In addition, tropical (but usually not temperate) cockroaches, millipedes, and scarab beetles are home to methanogenic bacteria.

A cow belches 50 L of methane daily. Even omnivores such as humans can be contributors via belches and flatulence. In the complex ecosystem of the gut, much of the methane produced by methanogens is scavenged by methane oxidizers and thus never sees the light of day. However, depending on diet and how balanced their digestive ecosystem is, animals may emit significant amounts of methane as one of the component gases of flatulence. To microbiologists, a cow is nothing more than a large fermentation vat for plant material. Thus, a contented, gas-emitting cow is a wonderful field mark for a healthy, fermenting community including methanogens.

DIGESTIVE TRACTS OF HUMANS

The intestines of humans are loaded with bacteria. The colon in particular is estimated to have more than 100 species of bacteria, mostly *Bacteroides* (see chapter 15) and gram-positives (chapter 12). These fermentative bacteria are exactly the sorts of compatriots with which methanogenic archaea thrive. However, not all human colons contain methanogens. In developed countries, 30 to 50% of people have methanogens in their colons, and therefore methane in their intestinal gas and sometimes even in their belches. In developing countries, the percent of methane emitters is greater. Children younger than 3 do not have methanogens but acquire them as their diets change to solid foods.

Tiny bubbles emitted from feces under water are good indicators of methanogenesis, as are feces that tend to float. The presence of methanogens seems to be in part a familial trait. However, families often have similar diets, so "familial trait" does not necessarily mean a genetic trait. It can be quite difficult to sort out whether a trait such as this is genetic or environmental. Twins both identical and fraternal have similar methanogenesis. Children of methane-emitting parents do too. It appears that one spouse can "give" methanogens to another. But is this due to similar diets, or to a passing around of actual bacteria and a genetic tendency to harbor them? A genetic influence is suggested by the fact that most Old World primates are methane emitters, whereas most New World primates are not, regardless of similarities of diet.

The normal passage of gas by humans ranges from 400 to 2,400 mL per day, composed mostly of nitrogen, oxygen, and carbon dioxide. These are odorless gases from swallowed air or diffused from the blood—or, in the case of carbon dioxide, from the many fermentative bacteria of the intestines. Also present may be hydrogen, methane, and trace sulfur compounds, the last of which make intestinal gas smell unpleasant (see chapter 9). Methane is odorless but flammable. Indigestible food in the diet, such as certain carbohydrates of legumes and lactose (in lactose-intolerant people), may increase the production of methane in those who have methanogens. However, these foods may also increase the activity of sulfide-producing bacteria, which out-compete the methanogens for the hydrogen gas that is necessary for both of their metabolisms. Therefore, if intestinal gas smells sulfurous, it is not likely to be rich in methane. Direct evidence for methane in flatulence is its flammability, which has on occasion caused explosions during electrosurgery of the colon, due to the right mix of methane, hydrogen gas, and oxygen—plus a spark. And a fraternity trick—not recommended here—is to attempt to light one's flatulence, with a little jet of blue flame indicating the presence of methane.

SEWAGE AND LANDFILLS

Methanogens thrive in the anoxic depths of waste waters and waste sediments such as those in home septic systems. Bubbles coming up from such places are likely to be good field marks for methanogens, just as they are in swampy sediments. Methane is one of the gases that is vented through the pipe that accompanies all toilet systems (to prevent explosive methane from accumulating in the plumbing). Therefore, consider the vent pipe to be a methanogen field mark! Most municipal sewage treatment plants are quite well aerated and rely on aerobic organisms to consume wastes. However, sludge generated from these plants may contain methane gas. Municipal sewage systems may also accumulate methane gas in their pipelines. Because this gas is explosive and dangerous in large quantities, workers in sewage systems are careful with sparks in enclosed spaces, yet there are still sometimes dangerous explosions of "sewer gas," or methane.

Enormous quantities of methane may be produced in those landfills that have deep layers of decaying garbage. Sometimes landfills are designed with outlet pipes to vent the methane, and sometimes the vent pipes are even set on fire to consume the methane, rather than releasing it into the air. If you see a landfill with flaming vent pipes, you are seeing a spectacular field mark of methanogens. A more economical use of methane from landfills is to channel it into a facility where it is burned to produce heat and electricity.

GEOTHERMAL SPRINGS

Methanogens that are not reliant on other organisms as a source of hydrogen are found in some hot springs, where hydrogen is available in geochemically produced anoxic waters. Any geothermal spring is likely to have methanogens present; however, evidence of gas emissions is not a truly reliable field mark because other gases are likely to be bubbling up as well, including sulfidic gases, which tend to dominate the senses.

VIEWING METHANOGENS UNDER A MICROSCOPE

Gather a bit of swampy sediment and examine it, keeping in mind that you are likely to be overwhelmed by an abundance of bacterial rods and cocci, including methanogens. There should be many fermenting bacteria in there as well, and with ordinary techniques, it is not easy to tell them apart from the methanogens. As it happens, though—and very conveniently for microbiologists with fluorescent microscopes—methanogens absorb violet and ultraviolet (UV) light and fluoresce (emit light). Using this technique, one can detect and enumerate methanogens with relative ease. (*Note*: Do *not* attempt to replicate a fluorescent scope at home by using UV lamps; you can seriously damage your skin and eyes.)

It was fluorescence microscopy that enabled researchers Thomas Fenchel and Bland Findlay to detect methanogens dwelling within or on the ciliates that inhabit anoxic sediments. If you encounter ciliates in sediment or in fresh or marine water that is anoxic, it is likely that you are looking at a symbiosis with methanogens. What seems to be going on in these symbioses is that the methanogens are being bathed in their favorite molecule, hydrogen gas, which is the waste product of fermentative metabolism in the ciliate. In turn, the ciliate is provided with a convenient means of getting rid of its waste product, which can be slow to diffuse from the cell, especially in large ciliates. By taking refuge in a ciliate, methanogens also avoid exposure to any oxygen that might be introduced to the environment, and also avoid competition with many other hydrogen scavengers, such as *Desulfovibrio*. If, by chance, you happen to have some cow rumen fluid to examine, or a freshly opened damp-wood termite such as *Reticulitermes* or *Zootermopsis*, you are also likely to see protists associated with methanogens (see chapters 12 and 16).

CULTURING METHANOGENS

The methanogens are anaerobes, poisoned by oxygen. Therefore, it is not an amateur project to cultivate methanogens on their own. You can, however, easily bring a bit of the swamp indoors by filling a jar with hay or dry leaves and topping it off with swamp water. In just a few days, methanogens will signal their presence with bubbles. Try giving the jar a shake or stir to release a champagne-like effervescence.

OTHER BACTERIAL GROUPS ASSOCIATED WITH METHANOGENS

Fermenters that produce hydrogen gas, carbon dioxide, and acetate are essential to the well-being of most methanogens. (The few exceptions are those methanogens that enjoy the anoxic waters of geothermal springs.) Thus, it is likely that when there are methanogens in swamps, other anoxic sediments, or digestive systems, there are also fermenters. Unfortunately, this does not mean that the presence of methanogens is a reliable field mark for any single bacterial group, since fermentation is a very common type of metabolism. However, methane oxidizers and methanol-using bacteria (see chapter 6 on alpha proteobacteria) are a specific group of bacteria that rely on methanogens for their food, methane. Where there are methanogens, then, there are likely to be methane oxidizers nearby in oxygen-rich sediments. Most methane escapes into the atmosphere and is promptly oxidized, but the methane that remains in sediments and waters is scavenged by methane oxidizers. The fact that we have even traces of methane in our atmosphere is a testament to how quickly and abundantly it is being produced all over the world.

SUMMARY: FIELD MARKS AND HABITATS OF METHANOGENS

- bubbles of methane in a stagnant, anoxic freshwater environment (e.g., swamps, bogs, sludge) coming up spontaneously or when stirred up
- flare-up of a lighter or candle when held in the vicinity of a bubble
- spontaneously lit "swamp gas," visible as "ghost lights"
- bubbles and even popping sounds given off by some aquatic plants
- "wet wood" of certain trees (which sometimes produces enough methane under pressure to make a flame)
- belches and flatulence of herbivores and omnivores, including humans (especially with no sulfury smell)
- feces that float or emit tiny bubbles that do not smell sulfury
- bubbles in a home septic system and vent pipes connected to toilets
- gas (sometimes explosive) in sewer pipes
- gas jets in large landfills (which burn off methane)

CHAPTER 3

Hyperthermophilic Archaea

The specific hot environments and field marks of the hyperthermophilic archaea are distinctive enough that you should be able to distinguish them from the ancient hyperthermophiles and *Thermus* species. In addition to favoring temperatures greater than 80°C (176°F), the hyperthermophilic archaea thrive in extremely low pH, often less than 3. (To get a full appreciation of hyperthermophilic diversity, see chapters 1 and 17 in accompaniment with this chapter.)

One specialist in hyperthermophiles, especially of the archaea, is the German scientist Karl O. Stetter, who makes his collections and observations in solfataric fields—terrestrial muds and waters heated by volcanic activity. These fields are rich in sulfur compounds and tend to be acidic as well as boiling hot. "It's like fire and brimstone," announces Stetter as he gingerly steps over the fragile, steaming crust of a solfataric field in the video *The Immortal Thread*, from the series *Secret of Life*. The surface crust of solfataric mud tends to be yellowish due to relatively oxidized iron sulfur compounds in contact with the atmosphere. Beneath the crust, the sulfurous muds are black, just as they are in other sulfur-rich environments such as estuaries and salt flats. The blackness indicates ferrous (iron) sulfides and related compounds formed in the absence of oxygen. It is that lack of oxygen, as well as the heat, that is favored by many bacteria of ancient lineage, such as the archaea and some of those described in chapters 1 and 17.

Many of the hyperthermophilic archaea are heterotrophic, anaerobic, and use sulfur compounds similarly to how we oxygen-breathing heterotrophs use oxygen (see chapter 5, on the sulfate-reducing bacteria, for more about this type of sulfur-breathing heterotrophy). Others are autotrophs, gaining energy from minerals (rather than light) to make food.

In addition to being hyperthermophilic, many archaea are also extreme acidophiles (lovers of acidic, or low-pH, conditions). They even contribute to the acidity of their environments by converting sulfide compounds to sulfuric acid. *Sulfolobus* is the genus most often isolated from the thermal areas described next.

Field Marks and Habitats of Hyperthermophilic Archaea

The best location in the United States—and one of the best worldwide—for observing the field marks of hyperthermophilic archaea is Yellowstone Park (table 3.1). (This park is more completely described in chapter 1.) At Yellowstone, some of the geothermal features are characterized by sufficiently high temperatures (greater than 80°C [176°F]) and amounts of sulfur, and low enough pH (e.g., less than 3) to harbor hyperthermophilic archaea. Look in particular for thick, simmering mud with a deep gray-black anaerobic zone and a top yellowish-rusty zone, indicating iron. The area should smell of sulfur. An iridescent, oily, glimmering layer on top of the sulfurous mud is sometimes a field mark for *Sulfolobus* and related genera. Muds at a full rolling boil, however, are not stable enough to

TABLE 3.1. Features of Yellowstone Park Likely to Have Hyperthermophilic Archaea

Feature	Specific field marks
Norris Geyser Basin	Most of the features are sufficiently hot to harbor thermophiles, but look for relatively stable muds and pools rather than geysers, which might disrupt the field marks
Porcelain Terrace Springs	Sulfurous, 95°C (203°F)
Mud Volcano Area	Most of the features are sufficiently hot, sulfurous, and acidic
Mud Volcano	Sulfurous, 84°C (184°F)
Black Dragon's Cauldron	Sulfurous, 88°C (191°F)

TABLE 3.2. Interpretations of Colors in Acidic Features

Color of water or sediment	Organism or mineral present
Yellow mud and water	Sulfur compounds
Greenish-yellow mud and water	Sulfur compounds, with a bluish tint from silica minerals
Red to reddish-brown deposits	Iron oxides: iron that was dissolved by acidic water and then deposited as "rust"
Bright, grassy green sediments	*Cyanidium*, an acid-tolerant green alga that can live at temperatures as high as 54°C (130°F)

show a surface sheen. *Note:* Boiling sulfurous muds are highly dangerous; you should not try to collect samples.

You must determine both pH and temperature to identify the bacteria correctly. Both measurements can be made using indirect methods. The first method is simply to trust the excellent documentation and signage on boiling mud features of Yellowstone as well as whatever information the park rangers can provide. For example, in the mud volcano area of Yellowstone, some pHs are identified as being as low as 0 to 2, and the temperatures often reach boiling point (see table 3.1).

Another method for determining acidity is to examine the acrid, sulfurous air. If you happen to have a piece of pH litmus paper with you, test the acidity of the vapors from the thermal features. Finally, mud itself is evidence of acidity because acidic waters dissolve clay minerals from the rocks below, producing a thick, muddy suspension. The more neutral to alkaline features described in chapters 1 and 17 have clear, often blue waters and white sediments.

The colorimeter table in chapter 1, by which the temperature of neutral to alkaline springs can be determined, is somewhat less useful for acidic features because these harbor a different assemblage of bacteria and minerals whose pigments are often obscured by the mud. The colors that can sometimes be seen in acidic features are listed in table 3.2 and are shown in plates 10 to 13.

Barring a trip to Yellowstone, you might be able to locate sulfurous, hot muds of correct temperature and pH in other locations—always using great caution around these dangerous areas.

Another archaean habitat is "self-heating" refuse piles of coal that are rich in sulfur; these may sometimes be found in coal mining areas. The high temperatures of these coal piles are a result of the metabolism of hyperthermophilic archaea and possibly other heat-loving bacteria. This process is similar to the self-heating of garden compost due to microbial metabolism. The acidic drainage from some iron-rich mines may also be due in part to archaea, although not necessarily thermophilic members. Thermal sulfur springs in the areas of deep-sea vents may also harbor hyperthermophiles, although these are for beyond the reach of most users of this field guide.

VIEWING HYPERTHERMOPHILIC ARCHAEA UNDER A MICROSCOPE

Even if you have a microscope, you should think twice before sampling boiling sulfurous mud. It is not worth putting yourself in danger, and your sample will quickly deteriorate under the air temperature conditions of your microscope. The mud will also thoroughly obscure what you might hope to see—a diversity of bacterial shapes. It is better to stay on safe trails, located at appropriate distances from cauldrons of fire and brimstone. Just as your field guide to the mammals of North America recommends that you stay a safe distance from bears, so this, your field guide to bacteria, recommends that you keep a safe distance from hyperthermophilic archaea in boiling mud.

SUBSURFACE BACTERIA

Some of the hyperthermophilic archaea, as well as other groups of bacteria, have subsurface, or "troglodyte" (cave-dwelling), relatives that live at a wide range of temperatures.

Subsurface bacteria that inhabit the cracks and crevices of rocks deep within the ground are both abundant and diverse. They are extremely slow-growing, reproducing over the course of months or years, and many are chemoautotrophs, using the energy in minerals to synthesize food. Some are thermophiles, living at the higher temperatures that characterize deep crust. In general, subsurface bacteria are inaccessible to amateur microbiologists as well as to most professionals. Their effects are subtle and, in most environments, invisible. For example, on the moist walls of a cave with abundant sulfides, faster-growing bacteria may predominate. Instead, if you can get access to cores of deep subsurface rocks, look for any evidence of minerals filling in the cracks and fractures. These minerals might well constitute field marks of slow-growing subsurface bacteria. However, identifying these bacteria is not an activity for amateur naturalists. Possible contenders include various archaebacteria, proteobacteria, and gram-positive bacteria, but the subtle field marks of the cracks are difficult to interpret.

SUMMARY: FIELD MARKS AND HABITATS OF HYPERTHERMOPHILIC ARCHAEA

- An iridescent, oily, or glimmering layer in hot springs and muds with temperatures of 80–110°C (176–230°F) and acidic, sulfurous, relatively stable conditions
- acidic, sulfur-rich waters or muds or soils of solfataric fields
- self-heating piles of sulfur-rich coal

CHAPTER 4

Halophiles

All bacteria—and in fact all living things—need some salt in their environment. Many bacteria require or tolerate fairly high levels of salt, often favoring marine environments. Very few bacteria can grow in extreme concentrations of salt (for example, ten times the salinity of seawater). Those that can, constitute a fairly cohesive phylogenetic group, the halophiles (literally, salt lovers). A branch of the archaea, the halophiles inhabit the salty crusts of marine salterns, or salt lakes. While there are other salt-tolerant bacteria that deserve the adjective "halophilic," in this field guide, *halophile* refers to a particular branch of the archaea. (Other halophilic bacteria are mentioned in this chapter too, as they often live side by side with the archaeal halophiles.)

Halophiles have a number of unique characteristics, presumably adapted to their extreme habitats. They vary in shape from coccoid to rods but often show pleomorphism (changes in shape), including changes to discs, triangles, or rectangle-like configurations. The most distinctive feature of these cells is the pink to reddish color of carotenoid pigments, which brightly tint the salt crystals on which they grow. (There are other pigments as well, which combine to give the distinctive color of the cell as a whole.)

What archaeal halophiles do to obtain food and energy is so strange that it defies characterization into any of the usual categories of bacterial metabolism. They are not strict heterotrophs or photosynthesizers, but a unique combination and variation of the two. Archaeal halophiles do not use chlorophyll, as all true photosynthesizers do, but rather a distinctive pink-red carotenoid pigment called bacteriorhodopsin (which looks very similar to the rhodopsins of vertebrate eyes). The bacteriorhodopsin is embedded in specialized regions of the membrane, called "purple membranes." When stimulated by light, the pigment acts as a pump, moving hydrogen ions (H^+; also called protons) across the membrane, storing them up as though storing charge in a battery. This isn't photosynthesis, because no food is being synthesized. Rather, this represents an auxiliary source of energy for the cells, a sort of back-up system to their usual metabolism, which is respiration, a fairly typical type of heterotrophy. An analogy for the halophilic archaea is a house powered mainly by the consumption of fuel for heat and light (heterotrophy), but with auxiliary solar panels on the roof (the purple membrane system).

FIELD MARKS AND HABITATS OF HALOPHILES

SALTERNS, SALINAS, OR EVAPORITES

A search for halophiles will lead you to some fascinating environments, well worth a detour on a family vacation. Head for a marine saltern (a place where marine water is evaporating) if you are near the West Coast, or a salt lake or other salty area if you are inland. Using an atlas and key words such as "salinas," "saline," and "salt" will help you find a likely area. Great Salt Lake of Utah is an obvious destination; however, other areas of the West, such as Texas, Wyoming, and New Mexico, have salty places, some of them remnants of the ancient seas that once covered parts of the North American continent. Use a guide book or phone the nearest tourist information center to be sure that you are headed for the right sort of place—a hypersaline environment with waters so saturated in salt that crystalline salt forms. Such an area may well be used commercially as a solar saltern or salt garden, such as the one in Grantsville, Utah. If the area is alkaline, as often indicated by names that include the word "soda," see chapter 9 for information about pink gamma proteobacteria that are likely to be present.

Having arrived, look for salt, either small amounts encrusting the shoreline of a body of water, or—far better—huge expanses of it, dazzling like a snow field, often incongruously under the hot sun. Look closer and note that all is not purely white but a gorgeous array of pinks and reds. In a commercial saltern, evaporation may be controlled by a series of holding ponds, each of which is a different shade of pink, producing an overall effect similar to a huge patch-work quilt. Because halophiles are sensitive to salt concentrations, more dilute holding ponds have fewer bacteria and therefore paler colors. Furthermore, halophiles seem to have some influence over the way that salts crystallize. Their presence helps increase the precipitation of sodium chloride over other salts.

Charles Darwin, in *Voyage of the Beagle*, described salt lakes in South America, noting the pink color as well as an underlying layer of black sediment. He strongly suspected that there was microscopic life in the brine, remarking that "parts of the lake seen from a short distance appeared of a reddish colour, and this perhaps was owing to some infusorial animalcula."

Get a close look at some of the salt crystals. These should look much larger than ordinary table salt, more like road salt or certain gourmet cooking salt. The crystals are likely to be pink owing to the presence of halophiles basking in the sun, carrying out their own peculiar version of photometabolism (fig. 4.1, plate 14).

If you have an opportunity to do so, wade out into a saline body of water—carefully. Be aware that the pink salt crust you are step-ping on may not be very thick and is often underlain by thick, black anaerobic mud, full of nonhalophilic sulfate-reducing bacteria (plate 15; see also chapter 9). There may also be a layer of salt-tolerant or halophilic cyanobacteria (chapter 13). Walk cautiously with bare feet to really enjoy the sensation of the warm, almost viscous salt water and the brittle, other-worldly landscape of crystals. The presence of bacteria enhances the precipitation of salts. A typical order of pre-cipitation is "calcareous lime" salts first, followed by halite (sodium chloride, or common salt) and then "bitter salts" of magnesium such as Epsom salts. Expect to see, here and there, crystals other than halite. If you're lucky, you may also see large, pink, flower-like "rose gypsum" (hydrous calcium sulfate, or $CaSO_4 2H_2O$), anhydrite (calcium sulfate, or $CaSO_4$), and dolomite (calcium magnesium carbonate, or $(CaMgCO_3)_2$).

SALT MINES

Salt mines, even deep ones, are locations where halophilic bacteria once lived, because they are the evaporated remnants of long-gone seas. One type of mineral deposit is called a salt "dome," which is

Figure 4.1. A commercial saltern in Mexico where the salt water of a lagoon is evaporated. The salt is being collected in coconut sacks.

essentially a mountain of salt left by an evaporating sea and then piled up by geological forces. In some arid places, brine from ancient salt deposits may leach up and leave salt crystals.

NUTRIENT-RICH RUNOFF

Another source of concentrated salt, and thus an opportunity for halophiles, is runoff from excessively fertilized fields. If the fertilizer salts end up in shallow ponds where the water can evaporate, the right sort of saline conditions may exist to support colorful shades of pink bacteria.

SALTY FOODS

Another field mark of the halophilic archaea is the red pigmentation of salted fish or the "red heat" condition of salted animal hides that are contaminated with halophiles. However, it is increasingly rare to use salting exclusively as a method of long-term preservation, and the halophiles do not thrive under refrigeration. Therefore, food preserved by both salting and refrigeration is not a reliable source for viewing halophiles.

Thai fish sauce, however, provides a way to enjoy the flavors of halophiles, which carry out fermentation of a mixture of salt and fish. Some gourmet cooking salts, harvested as large moist crystals from salterns, are also products of halophilic activities: halophiles seem to encourage a sequence of precipitation of sodium chloride that yields the best crystals. The colors of some gourmet salts are indicative of their bacterial influences. Pink tints are from the halophiles. Gray or black salt, sometimes smelling sharply of sulfur, was in contact with sulfate-reducing bacteria (chapter 9). However, note that some exotic salts are deliberately mixed with colored substances, such as red salt of Hawaii with iron-rich red clay.

The Swedish fermented herring *surstromming* (sour herring) is an extreme example of food enhanced by halophilic archaea. In the canned version, odorous gases produced by the microbes cause the can to bulge disconcertingly. The flavor of the fish is so extraordinary that very little is exported from Sweden except to Swedish expatriates who have grown up appreciating this delicacy. The same sort of bacterial activity is also featured in fermented puffer fish ovaries, a gourmet item in Japan. (For more on fermented fish, see chapter 10.)

VIEWING HALOPHILES UNDER A MICROSCOPE

Collect some water that is in close contact with pink salt crystals, and take a look. Halophilic archaea should appear in many shapes; however, geometric shapes—squares and triangles—are the best indicators of halophiles. Unfortunately, these shapes are not common. The salt-loving nature of these cells may be demonstrated by dropping them on pure water, which should cause the cells to burst. In addition to seeing bacteria of various shapes, you may also see the unusual eukaryotic red-pigmented "green" alga *Dunaliella*. There may also be salt-tolerant cyanobacteria in the sample (chapter 13).

CULTURING HALOPHILES

You may be able to collect some salt containing halophiles, along with the underlying mud. Collect this in a jar, and maintain it in a sunny window. Water the mix sparingly, enough to keep water above the crystals.

SUMMARY: FIELD MARKS AND HABITATS OF HALOPHILES

- pink, orange, or red pigments on salt crystals in marine salterns, salt lakes, salt crusts, or commercial salt evaporation sites*
- pink-tinted runoff from fertilized fields
- food products such as some large-crystal salts, Thai fish sauce, spoiled salt meats at room temperature, and fish with pink-red discoloration

* *Note:* the eukaryotic alga *Dunaliella* also contributes to the pink color in some salty environments. If a salt lake or crust is known to be alkaline, sometimes indicated by the word *soda*, a pink color might indicate gamma proteobacteria (see chap. 9).

CHAPTER 5

Green Sulfur Bacteria

The green sulfur bacteria are one of many groups of bacteria that take advantage of the wonders of photosynthesis to make their own food. The other photosynthetic bacteria are the green nonsulfurs, the cyanobacteria, some of the proteobacteria, and the heliobacteria. (The halophiles also collect light energy, but the process does not involve chlorophyll and food production and thus by strict definition is not photosynthesis.)

Metabolism in Green Sulfurs

The common name of green sulfurs summarizes what they are: *green* refers to their bacterial version of chlorophyll, an essential pigment for all photosynthesizers; *sulfur* means that they use hydrogen sulfide (H_2S, the gas that smells like rotten eggs) as a source of hydrogen. Most other photosynthesizers (cyanobacteria and all green plants) use water as a source of hydrogen. Compare these two simplified reactions for photosynthesis:

$$Green\ sulfur\ bacteria$$
$$CO_2 + H_2S + light \rightarrow C_6H_{12}O_6 + S$$
$$(sugar)$$

$$Cyanobacteria\ and\ all\ plants\ and\ algae$$
$$CO_2 + H_2O + light \rightarrow C_6H_{12}O_6 + O_2$$
$$(sugar)$$

Note that these groups differ not only in their source of hydrogen but also in their waste products. Cyanobacteria, green plants, and

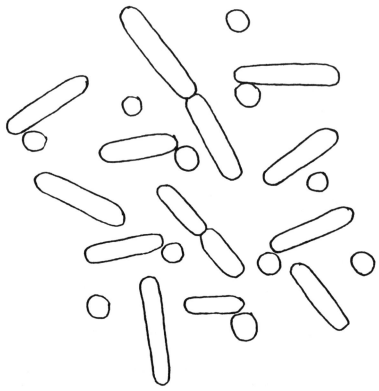

FIGURE 5.1. Green sulfur bacteria (shown here as rods) sometimes accumulate bright yellow-white globules of sulfur (shown as spheres). Sulfur is a waste product of photosynthesis.

algae are the source of nearly all the oxygen in our atmosphere, released as the waste product of photosynthesis. In contrast, green sulfur bacteria produce elemental sulfur, a yellowish solid that is not so easily discarded. Globules of sulfur often remain attached to the outsides of the cells, although in some environments the sulfur is converted to the more soluble sulfate (SO_4), and thus the bacteria are freed from carrying around globules of their own wastes (fig. 5.1).

Other organisms consider the sulfurous waste products of green sulfurs to be an important source of materials for their own metabolism. Sulfate-reducing bacteria (of the delta proteobacteria) "breathe" sulfate, in a process similar to how we use oxygen. Just as we oxygen-dependent organisms form tight interdependent ecological systems with our sources of oxygen (and food), so do the various sulfur-dependent organisms form tightly interwoven communities, called "sulfureta" (singular: sulfuretum). The field marks of these organisms are often quite distinctive, making the green sulfurs and other organisms of the sulfuretum fairly easy to identify. (For more on sulfureta, see chapter 9.)

FIELD MARKS AND HABITATS OF GREEN SULFURS

Wherever there is a rich source of sulfur compounds in the photic zone, there is a good possibility of seeing green sulfurs. The trick is to be able to distinguish them from the other green (or blue-green) bacteria, such as cyanobacteria (chapter 13), and from the larger green cells of eukaryotic algae, which can predominate in some aquatic systems. Essential environmental criteria for green sulfurs include sulfur (in the form of smelly hydrogen sulfide), a lack of oxygen, and the presence of light. A general awareness of the other members of a typical sulfuretum can also help in identification; therefore, consult chapter 9 about some of the important proteobacteria of the sulfuretum, including purple sulfur bacteria, *Beggiatoa*, and sulfate reducers. In sediments and waters, green sulfurs often form a green layer surrounded by the grays, whites, and pinks of the various proteobacteria, like the layers of a Napoleon pastry (plate 16).

By the way, the bioluminescent deep sea fish *Malocosteus* makes extraordinary use of the chlorophyll of green sulfurs, concentrating and modifying it to use as a pigment for visualizing red light (see chapter 9 for more detail).

MARINE WATERS

Marine estuaries and mudflats, especially those with a sulfur smell, are good places to look for green sulfurs. Look for relatively undisturbed sediments, either under shallow water or emerged but still damp. Expect to see a top layer of black-green to blue-green cyanobacteria. If you gently scrape away this layer, you may see a pink layer just underneath; these are purple sulfur bacteria (plate 17). In some very sulfury environments, the pink layer may be right at the surface. A white layer of *Beggiatoa* or related bacteria may or may not be present. Alternatively, you can cut down through the sediment with a knife or spatula to see if there are any thin colored layers.

A green layer of green sulfur bacteria below the pink-purple layer may or may not be present. Green sulfurs like a lot of hydrogen sulfide and are most likely to be found next to an underlying deep black layer of sulfate-reducing delta proteobacteria, which are the source of the hydrogen sulfide. If your sediment is not quite rich enough in hydrogen sulfide and is a little too oxidized, you may not see this green layer at all. Keep trying different places. The more stagnant the water, the better. Also, like many photosynthetic bacteria, green sulfurs sometimes display not their green colors but rather their yellow-brown accessory pigments, which can make them more difficult to identify.

FRESH WATERS

Layering similar to that described in the previous section can occur in stagnant freshwater ponds, lakes, or waste stabilization ponds. However, the pink and green layers are likely to be too deep to see without special methods. Field microbiologists sometimes pump water from a particular depth in order to collect bacteria from that layer.

SULFUR SPRINGS

Sediments in close contact with sulfur springs (cool to warm but not hot) may also be good places to look for a layer of green sulfurs, situated close to the source of sulfur yet still in the photic zone. Dim light is sufficient for green sulfurs—and indeed necessary, lest they be outcompeted by other photosynthesizers (plate 18).

Whatever greenish layer you see below a pink layer is extremely unlikely to be cyanobacteria, which thrive in more light. The photons that make it through the top layer of cyanobacteria and the next layer of pink are apparently sufficient only to support green sulfurs.

VIEWING GREEN SULFURS UNDER A MICROSCOPE

Remove a bit of the suspected green layer, put some on a slide, and pick out any tiny grains of sand before applying a coverslip. If you

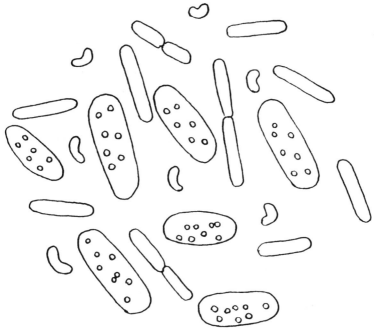

FIGURE 5.2.   Purple sulfur and green sulfur bacteria can often be found in the same sample examined under a microscope. The green sulfur bacteria may or may not have visible sulfur granules. The ones shown here are small rods and comma (vibrio) shapes without sulfur granules. The purple sulfur bacteria may appear to be scurrying about like little pink mice. Here they are shown as large rods with sulfur granules inside.

see beautiful long blue-green filaments, you may have accidentally picked up a layer of cyanobacteria. Green sulfurs are smaller green cells that come in a variety of shapes—ovoids, rods, and spirals—and may have tiny bright globules of white-yellow sulfur on their outsides. You may notice some purple sulfur bacteria too (chapter 9), sometimes looking quite motile among the relatively nonmotile green sulfurs (fig. 5.2).

The cells of the genus *Chlorobium* (the best known of the green sulfurs) may be coccoid or rod shaped, ranging from straight to curved or spiral. Other green sulfurs have a similar range of shapes. In some species of *Pelodictyon*, the cells may form aggregates or even

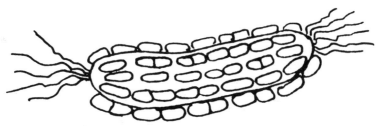

Figure 5.3.   Some green sulfur bacteria are found in consortia (symbioses). Here, *Chlorochromatium* is shown as a colorless, rod-shaped bacterium with tufts of flagella at each end and a covering of much smaller, rod-shaped green sulfur bacteria.

net-like arrays. Some species of *Prosthecochloris*, a common marine green sulfur, have prosthecae, which are long stemlike appendages.

Most intriguing are those freshwater green sulfurs that have formed symbioses (also called consortia) with other bacteria. These consortia have been given the genus names *Chlorochromatium* and *Pelochromatium*. Typically the association consists of barrel-shaped green sulfurs arranged in 5 to 6 rows around a central colorless bacterium. If you are fortunate, you might see one of these under a microscope (fig. 5.3). In one consortium, *Chloroplana*, the cells are arranged in flat sheets, alternating green and colorless. The identity of the colorless cell is unknown because it cannot be cultured outside of its symbiosis. It is hypothesized to be a sulfate-reducing proteobacterium, which makes good sense since the two are almost always found in close proximity in sulfureta. Green sulfurs generate sulfur and sulfate compounds while sulfate-reducing bacteria use those compounds. The sulfate-reducing bacterium is often motile, due to the use of a flagellum. This motility may confer an advantage by helping the consortium to situate itself in the photic zone when there is little or no oxygen yet plenty of hydrogen sulfide (chapter 9).

CULTURING GREEN SULFURS

Green sulfurs can be cultured fairly easily. See the instructions for making a Winogradsky column in appendix A and chapter 9; this is a good way to get a look at the differences between the cyanobacteria in the top layer and the green sulfurs in one of the layers below. Look for the green sulfurs right next to the black layer of sulfate reducers.

SUMMARY: FIELD MARKS AND HABITATS
OF GREEN SULFUR BACTERIA

- green layer in the photic zone (top few mm) of sulfide-smelling marine sediments; for most accurate identification, look for the layer just under a thin layer of pink-purple bacteria and above the black, sulfide-rich sediment
- green layer deep in a stagnant pond; can usually be seen only by selectively sampling water at different depths while taking care not to disrupt the layers
- green layer in close contact with a sulfide-rich spring (but not a thermal spring), preferably just below a layer of pink-purple bacteria

---

Introduction to
the Proteobacteria

The proteobacteria (also called purple bacteria, after some of their pigmented members) are a highly diverse and extremely successful group. All proteobacteria are gram-negative, which means that they do not become stained with crystal violet (Gram's stain), unlike the gram-positives (see chapters 10–12). This may seem like a trivial matter, and in fact the staining procedure itself is quite trivial; however, all bacteria (except the archaea) are essentially divided into one of two large groups based on this stain. Thus, Gram-staining has become one of the starting points for identification of almost any new microbe (see the introduction to the gram-positive bacteria for more on the history of the Gram stain).

The new phylogeny of organisms, based on DNA sequences, has revealed not only that all bacteria but the archaea can be divided according to their Gram-staining properties (which are influenced by their cell wall type) but also that the gram-positives are on one big branch of the bacterial tree and almost all well-known gram-negatives are on another big branch. This leaves about nine other major branches on the non-archaeal side of the tree, all of which turn out to be gram-negatives as well. The archaea have different cell wall structures and are not classified as gram-negative or -positive.

The intrepid field microbiologist Lynn Margulis once said that most organisms are gram-negative rods. They are analogous to the indistinctive "little brown birds," or LBBs, of ornithologists. Indeed, the ubiquity of this group means that any scoop of soil or sediment, any bucket of water (no matter what its salinity), any surface or body cavity of any organism is almost guaranteed to contain gram-negatives. And chances are that almost any gram-negative bacterium you happen to encounter is a proteobacterium (though cyanobacteria, which are also gram-negative, are extremely abundant in some environments).

The proteobacteria are divided into five subgroups named by the Greek letters alpha, beta, gamma, delta, and epsilon, based on their DNA sequences. This is a bit inconvenient for naturalists interested in memorizing significant nomenclature, particularly since the Greek letters indicate nothing in themselves. This starkly objective nomenclature reflects the great diversity of bacteria within each subgroup; no descriptive name, newly coined or revived from the literature, would suffice to summarize each of the five groups. Therefore, memorize the five proteobacterial groups and join the ranks of professional microbiologists in referring to this or that bacterium by its nickname—as in, "an alpha proteo" or "a beta."

The following four chapters describe field marks of the alphas, betas, gammas, and deltas. Chapter 8 focuses on the gammas and deltas that form fascinating, complex communities in a diversity of sulfur-rich environments. That chapter features macroscopic field marks of bacteria at their best—highly colorful and odoriferous. A trip to a sulfur-rich environment is highly recommended.

The epsilon proteobacteria are a small group consisting mostly of a few commensals, opportunists, and pathogens. They do not seem to have reliable field marks within the nonpathogenic purview of this field guide. One genus, *Helicobacter*, is now thought to be a likely cause of stomach ulcers. *Helicobacter* appears to be part of the normal microbes of the digestive tract and one of the few that can brave the harsh acids and tidal wave–like mixing of the stomach. What triggers the bacteria to become invasive and to cause ulcers is not well understood.

CHAPTER 6

Alpha Proteobacteria

The alphas are everywhere. Some of them reveal themselves with field marks, but many others are not easily detected. Swampy places and stagnant bodies of water are good places to find alphas. Swellings on the roots of some plants are associated with symbiotic, nitrogen-fixing alphas, while swellings on plant stems are associated with other alphas. A few alphas join the illustrious ranks of the gram-positives as food enhancers. Some make their own magnets with which they orient themselves. The cryptic alphas, those without field marks, include many soil bacteria and a few opportunists and pathogens such as *Rickettsia*.

Although some would not refer to them as bacteria, the mitochondria, cellular organelles that carry out the chemical process of respiration for us and other oxygen-using eukaryotes, were at one time free-living alpha proteobacteria. These bacteria became internal symbionts of complex cells about $2^1/_2$ billion years ago and evolved into present-day mitochondria; you can even think of yourself as a walking, breathing multicellular community.

FORMER ALPHAS WITHIN US: MITOCHONDRIA

Two-and-a-half billion years ago, it was a bacterial world. The plantless landscape was covered with slimes, scums, crusts, and mats of bacteria, some of them beautifully colored in reds, oranges, yellows, purples, pinks, or greens. If we, traveling in time, somehow found ourselves surveying this world, we would have to wear masks connected to oxygen tanks. There was very little oxygen in that ancient atmosphere: less than 1%, compared with about 20% in our modern environment. It was a paradise for anaerobes—the air smelled of their volatile waste products, and the sediments and waters bubbled merrily due to their activities. Alas, oxygen, a poisonous gas, was slowly seeping into this thriving microbial biosphere.

Around the time of the beginning of life (4 billion years ago), the atmosphere was virtually free of oxygen. This was a good thing, in fact, because most of the molecules that made up the organisms living at that time were extremely susceptible to being oxidized and destroyed. Many theoreticians suggest that life could not have begun if oxygen had been present. However, about $3^1/_2$ billion years ago, a new type of photosynthesizer, cyanobacteria, evolved and began to release oxygen as a waste product. Oxygen accumulated slowly, first oxidizing many of the sediments and waters and then building up in the atmosphere. Bacteria everywhere evolved and adapted (or went extinct). Some retreated deep into sediments and waters—still anaerobic zones to this day. Others evolved simple chemical mechanisms by which oxygen could be detoxified, usually allowing some specific molecule in the cell to become oxidized and thus removing oxygen from contact with all other molecules in the cell. (Reactions such as metal oxidation and bioluminescence have been hypothesized to be oxygen detoxification mechanisms; these are described later in this chapter.)

Still other bacteria, many of them proteobacteria, evolved oxygen respiration, in which oxygen could be both detoxified and put to good use at the end of a complex pathway for acquiring energy from food. A good analogy for this type of pathway is a waterfall turning a waterwheel. As long as the water keeps falling, useful energy can be collected by the waterwheel. However, if the pool of water at the bottom is not drained properly, it fills until the waterfall is inundated and all water flow stops. In respiration, oxygen is like the drain (or sink) for a flow of electrons. Without oxygen, the flow would slow down or stop. In this pathway, oxygen is converted to water (H_2O) and thus is detoxified to become part of a very useful compound.

In equation form, respiration does not look very different from fermentation. Both are means by which the energy from food molecules is captured and stored in the form of ATP (adenosine triphosphate) for later use in the cell. Both are variations of heterotrophy. However, the extra steps in respiration make it a much more effective process than fermentation. Compare these abbreviated representations:

Fermentation

Food molecule → Smaller molecule + CO_2 + ATP
(such as sugar)   (such as alcohol   (universal energy-
                  or acid)   carrying molecule)

Respiration

Food molecule + O_2 → CO_2 + H_2O + ATP
(such as sugar)   (universal energy-
                                     carrying molecule)

Note that the smaller molecule generated in fermentation—usually an alcohol or acid—is a waste product, potentially hazardous to the cell if it accumulates. Carbon dioxide, produced in respiration, poses less of a problem because it diffuses away more easily. Furthermore, the number of ATPs generated by respiration is about 10 times that of fermentation: about 36 molecules for every sugar (glucose) molecule consumed, rather than 3. ATPs are universal storage molecules for energy in the cell. They are roughly analogous to storage batteries. A cell requires thousands of them every minute, or even every second, depending on how active it is. Oxygen-using respiration was such a wonderful evolutionary innovation for producing ATP that the proteobacteria became a great success, outcompeting many other bacteria as oxygen increased in the atmosphere.

About $2\frac{1}{2}$ billion years ago, in a hot, acidic sulfur spring, dwelt some hyperthermophilic archaea, perhaps *Sulfolobus* or *Thermophilus* (chapter 3). Some of these archaea joined in symbioses with alpha proteobacteria similar to *Paracoccus*, and together became a new type of cell. This cell was complex and was capable of acquiring energy and coping with oxygen even better than either of the individual cells was. This was the probable origin of complex, eukaryotic cells that evolved into protists, fungi, plants, and animals—most of them highly aerobic and quite dependent on the 20% oxygen we now breathe.

We can consider ourselves and all animals (as well as most protists and all fungi and plants) as living field marks of those long ago alpha proteobacterial symbionts, as those symbionts are now our mitochondria. Our enormous size, extraordinary level of activity (in spite of gravity), and quick muscles and nervous systems are all testament to the efficiency of mitochondria in gathering energy. And just hold your breath for a while to remind yourself how obligately dependent we are on oxygen—which, although it continues to be a poisonous gas, has been put to excellent use by our symbiotic bacteria.

VIEWING MITOCHONDRIA UNDER A MICROSCOPE

Mitochondria (our symbiotic alpha proteos) have undergone many modifications in their $2^1/_2$ billion years of evolution within our cells. They are tiny, lack cell walls, and are often rather amorphous. They are difficult to visualize and distinguish from other cell parts under a microscope. Even professionally prepared and stained slides used to teach cell biology do not display mitochondria very well. An exception is the following project for visualizing mitochondria, which can be done in a classroom or at home by an ambitious amateur.

Acquire a stain called Janus Green from a biological supply house such as Ward's, Carolina, or Connecticut Valley. Obtain some paramecia (slipper shaped, single-celled protists) from that same supply house, or collect some in pond water. Prepare a stock solution of the stain by dissolving 1 gram of stain in 100 mL of 100% ethanol. Mix 2 mL of the stain with 28 mL of water, and place the solution in a jar wide enough to insert a microscope slide. Also have on hand a jar of alcohol (rubbing alcohol—isopropanol, for example). Cautiously heat a slide by holding it with forceps, dipping it into alcohol, and setting it on fire. Hold the slide so that the flaming alcohol does not drip onto your hand. Quickly stop any inadvertent alcohol fires; if flames appear in your alcohol jar, cover it or add a lid. When the flaming slide has burned out, dip it (still hot) into the Janus Green stain. Let the slide drain, then place the slide on its edge on a piece of paper towel. A pale green film should coat both sides.

Take a drop of the paramecium culture and place it on the slide. Add a coverslip and locate a paramecium under the microscope using a 10× lens at first, but later if possible a 20× or 40× lens. The paramecium should be full of green-blue rounded compartments, especially in its posterior end. These are mitochondria. The color is due to the oxidation of the dye by mitochondrial activity. In the absence of oxidation, the dye is colorless. A larger, oval nucleus as well as round food vacuoles may be visible in the center of the cell. Mitochondria are about a tenth the size of either of those structures.

AGROBACTERIUM AND OTHER GALL PRODUCERS

Plants suffer from wilts, clogged vascular tissues, necroses, blights, spots, rots, and galls (tumors). Fungi are major causal agents of plant diseases, as are nematodes, insects (sometimes indirectly causing infections through wounds), viruses, and bacteria. The groups of bacteria most likely to be associated with diseases of plants are in the proteobacteria. These include some of the many pseudomonads (with representatives in the alphas, betas, and gammas), *Erwinia* (a gamma), and *Agrobacterium* (an alpha). In addition, some of the gram-positives, including actinomycetes, are implicated in plant diseases. The diagnosis of plant diseases is usually quite complex, but *Agrobacterium* is an exception: it makes conspicuous galls on a wide variety of herbaceous plants, shrubs, and trees. Although this guide does not cover pathogens, *Agrobacterium* is included because the results of its activity are a fairly reliable field mark, and the growths are often benign.

BIOLOGY OF *AGROBACTERIUM*

Agrobacterium enters a plant via a wound, made by a nematode or insect, or by a grafting procedure or other injury. Once inside, *Agrobacterium* inserts into the plant cells a special set of genes, some of which are used by the plant as if they were its own genes. Some of these genes code for mimics of growth hormones produced by the plant; other genes code for food molecules that are useful to *Agrobacterium* but not to the plant. The plant cells begin to proliferate, producing a rich source of food for the bacteria. The resulting tumor or cancer-like growth is called a gall. In spite of the analogy to cancer, these bacteria-induced cell proliferations are not usually harmful.

Laboratory strains of *Agrobacterium* have become quite popular tools for genetic engineering. Molecular biologists attach genes of interest to the genes of *Agrobacterium*, which inserts these new genes into the plant.

FIELD MARKS AND HABITATS OF *AGROBACTERIUM*

Outside the laboratory, look for agrobacterial galls on shrubs or trees that have been grafted, a common procedure with some ornamentals and fruit trees. The galls should be at the bases of plants near the soil, where the original graft was made. Look especially at grapes, raspberries, roses, fruit and nut trees, willow, honeysuckle, euonymus, daisies, asters, chrysanthemums, and gypsophila. The galls in these plants are round, rough-textured swellings, up to several inches in diameter. Some grafted apple trees produce a gall that has rootlike structures protruding from it, suggesting that the hormone mimics have confused the process of cell differentiation in the plant.

Douglas fir and big cone spruce have their own species of *Agrobacterium*, which is spread by spruce gall aphids throughout the twigs and branches. However, not all galls are formed by *Agrobacterium*. Fungi, nematodes, mites, and insects are all important gall producers, with or without the help of bacteria. For example, the wonderful round apple galls of oak trees and the woolly galls on oak leaves are both made by tiny wasps. Sometimes a life cycle stage of the wasp may be found within.

Large galls (sometimes hundreds of pounds) are called burls or burrs. They can be found on the trunks of old trees and are in demand by some cabinetmakers for ornamental woodwork (figs. 6.1a, b). Trees with a natural susceptibility to making burls (such as elms) are considered interesting features in ornamental gardens and parks. Sometimes fantastic gnarled and knotted burls sprout shoots (known as epicormic or adventitious shoots or water sprouts), giving the tree a wild hairy, feathered, or fingered appearance. Colloquial names for these sprouting burls include "cats' heads" or "witches'-broom." They impart an interesting fairy-tale look, as depicted in illustrations by Arthur Rackham. Note that "witches'-broom" is most often used as a general name for multiple adventitious sprouts that may also be caused by some fungi and other bacteria. Therefore multiple sprouts alone are not a reliable indicator of agrobacterial activity.

FIGURE 6.1a. Agrobacterial swellings of a tree.

Some injuries to trees result in the formation of gall-like structures with or without shoots because the natural tendency is for trees to heal wounds by covering them over with a proliferation of wood and bark. Often, but not always, such injuries can introduce *Agrobacterium* and other microbes, making determination of a cause difficult.

In addition to injuries by insects and fungi, there are pruning-type injuries of trees sometimes introduced by humans to get particular aesthetic or practical effects. These including coppicing (cutting the main stem at ground level) and pollarding (cutting the main stem up high). Both can result in knots, gnarls, and lots of sprouts, some

FIGURE **6.1b.** Agrobacterial swellings depicted by
the illustrator of children's books Arthur Rackham.

of which may be enhanced by *Agrobacterium*. In some ornamental
gardens the trees are deliberately pruned to create a pollard effect.
Coppicing and pollarding were also once considered good ways to
get lots of small firewood without destroying the whole tree.

VIEWING *AGROBACTERIUM* UNDER A MICROSCOPE

Microscopic preparations of tough plant material are usually not
very successful, and the *Agrobacterium* cells tend to be highly
modified to the point of being unrecognizable as bacteria.

CULTURING *AGROBACTERIUM*

There are no reliable methods for amateur microbiologists to culture *Agrobacterium*.

RHIZOBIUM AND OTHER NITROGEN FIXERS

Many of the nitrogen fixers, and especially those that form distinctive symbioses with plants, are alpha proteobacteria. *Rhizobium* is a soil nitrogen fixer, famous for its symbioses with legumes. *Azospirillum* and its relatives colonize the surfaces of plant roots. Other nitrogen-fixing alphas enrich the soil with their activities. Probably all terrestrial plants are dependent on nitrogen fixers, whether in the soil or as symbionts, because nitrogen in a useable form such as ammonia or nitrate is in high demand. All organisms need it to build proteins, DNA, RNA, and other crucial molecules. Heterotrophs get their nitrogen by ingesting other organisms—and with them, their protein, DNA, and RNA. Plants take in their nitrogen as nitrate (NO_4) or ammonia (NH_3) via absorptive roots in the soil. If soil nitrogen were not replenished constantly, a growing plant would soon deplete whatever nitrogen there was in the immediate area. This is where nitrogen fixers become important. They take in gaseous nitrogen (N_2) from the atmosphere, where it is abundant (79% of the atmosphere) but not directly useful to organisms. They then convert it to ammonia, which is a form of nitrogen that can be used by other organisms. The reaction consumes a considerable amount of energy and requires an anaerobic environment.

Sometimes in an agricultural setting, soil nitrogen becomes depleted because too many plants are being made to grow too quickly in a small area. That is where fertilizer comes in; the nitrate in it temporarily replenishes the soil. However, it is often better to allow fields to lie fallow periodically to be replenished by bacteria. A common practice in fallow fields is to sow a legume that comes with its own symbiotic nitrogen fixers.

FIELD MARKS AND HABITATS OF *RHIZOBIUM*

Look for *Rhizobium* and related genera by looking for nodules on the roots of legumes. In the garden, these are peas and beans of all sorts.

On the lawn or in fields, look for clovers and vetches. Among the trees look for locust species. In the southern United States look for Kudzu vines. The tropics are home to many other legumes, such as certain species of *Sesbania*, which have nodules in their stems. Identify legumes (plants in the family Leguminosae) with an appropriate botanical guide book.

To see a field mark of nitrogen-fixing bacteria, simply dig up a legume and carefully rinse the soil from its roots (figs. 6.2–6.4). The nodules, spherical or oblong swellings, one to several millimeters in diameter, may be found as outgrowths of the roots. If you pick off a nodule and crush it between your fingers, you will see that the inside is pink. This coloration is due to a type of hemoglobin (leg hemoglobin) made by the bacteria and host together. In animals, hemoglobin binds to oxygen readily and is used in blood to carry oxygen from one place to another. In the plant, leg hemoglobin is used to bind oxygen and prevent it from interfering with the nitrogen-fixing reaction.

Plants that are in sandy soils, such as in deserts or dunes, often have relationships with alpha proteos and other nitrogen-fixing bac-

FIGURE 6.2. A legume such as red clover is likely to have nitrogen-fixing nodules on its roots.

FIGURE 6.3. A legume removed from the soil and rinsed.

teria. These plants include beach grass and some cacti; however, the field marks of such associations are not as visible as in the legumes. In or near some plants, other symbionts such as nitrogen-fixing cyanobacteria may be present (see chapter 14).

VIEWING NITROGEN FIXERS UNDER A MICROSCOPE

You can place a crushed nodule under your scope; however, it will be difficult to see the highly modified bacterial symbionts among the large, tough plant cells.

FIGURE 6.4. Nodules of nitrogen-fixing bacteria. Try crushing a nodule to see the pink color of leg hemoglobin inside.

CULTURING NITROGEN FIXERS

The first step is to grow lots of legumes in your garden. You can also buy packets of freeze-dried nitrogen fixers at a garden supply store and use this as a soil additive to give the legumes a head start. Usually plants must recruit nitrogen fixers from the surrounding soil, so if the soil is very depleted, an inoculation of nitrogen-fixing bacteria helps.

WOLBACHIA: A SYMBIONT OF INSECTS AND OTHER INVERTEBRATES

The alpha proteobacterium *Wolbachia* is a widespread, well-integrated symbiont of many invertebrates, perhaps even a majority of them. There appears to be no serious harm caused by *Wolbachia*, which means that it would be essentially undetectable at a macroscopic level were it not for the ability of some strains to skew the sex ratios of some insects in favor of females, by feminizing the males or killing them outright. Yes, it does sound a bit pathogenic. However, populations of the infected insects appear to be thriving, and *Wolbachia* itself enhances its own reproductive success: it is transmitted only through eggs (not sperm) and therefore benefits from an abundance of females in an insect population.

Consider the field mark of *Wolbachia* to be the skewed sex ratio (in favor of females) in some insects and perhaps other invertebrates. Note, however, that skewed sex ratios are quite common among insects (e.g., many bees and wasps), and are not all caused by *Wolbachia*.

VIEWING *WOLBACHIA* UNDER A MICROSCOPE

Wolbachia is so modified in its morphology as an obligate symbiont that it is not easily detected with microscopy, nor is it easily cultured.

METHANOTROPHS

In any anaerobic environment, there is almost certain to be some sort of methane cycle. Evidence for such a cycle is most likely to be observed in freshwater, swampy areas with lots of anaerobic sediments, in the anaerobic parts of animal digestive systems, or in landfills or water-treatment areas (methane-producing bacteria are outcompeted by sulfur bacteria in marine habitats). The main *producers* of methane are the methanogenic archaea (see chapter 2). Some of the *consumers* or oxidizers of methane are alpha proteobacteria, including the genera *Methylobacteria*, *Methylococcus*, *Xanthobacter*, and their relatives. (Note that *Xanthobacter* is easily confused with *Xanthomonas*, a plant pathogen of the gamma proteobacteria. *Xantho-* means yellow and is a common scientific prefix.) The methane-using (methanotrophic) bacteria carry out a type of autotrophy in that they produce food molecules from a simple starting compound: methane. However, they are not photoautotrophs; the energy for this type of autotrophy is chemical, not photic.

Some of the methanotrophs can perform other exotic metabolisms in addition to methane oxidation. For example, *Xanthobacter* is one of the "knallgas" bacteria, capable of an extraordinary type of chemoautotrophy (see table 7.1 for characteristics of chemoautotrophs). *Knall* is German for "crack" or "bang" as from fireworks, and knall gas is an explosive mixture of hydrogen and oxygen. *Xanthobacter* and similar bacteria oxidize hydrogen gas that accumulates in some methane-rich environments:

$$2H_2 + O_2 \rightarrow 2H_2O$$

The bacteria manage to control this reaction enough to gain useful energy from it and produce their own food as autotrophs.

FIELD MARKS AND HABITATS OF METHANOTROPHS

None of the methane-using alphas are easy to detect directly. Many are copious slime producers, although this alone is not a reliable field mark. However, methanogens (producers of methane) can be observed by the release of bubbles of methane from anaerobic, swampy sediments, especially when stirred up with a stick. And if there are methanogens present, then various methane-using bacteria are likely to be present too. These can be found in the aerobic zones just above the anaerobic sediment or water layer. Some deep-sea animals that have symbiotic sulfur bacteria (see chapter 9) also have symbiotic methanogens and methanotrophs.

CULTURING METHANOTROPHS

There are no simple methods for amateur microbiologists to culture methane-using or knallgas bacteria or to distinguish their morphologies under the microscope. Indeed, culturing knallgas bacteria is considered to be a dangerous undertaking as it requires explosive mixtures of hydrogen and oxygen in the lab. Technical instructions are often replete with emphatic warnings followed by exclamation points! Less than optimal gas mixtures (from the point of view of the bacteria) are often recommended for the safety of the scientists.

SEWAGE CONSUMERS

Many bacteria, protists, fungi, and small animals (e.g., nematodes) enjoy the nutrient-rich waters of sewage. Their activities are essential for breaking down the solids in sewage and for purifying the water, and sewage-treatment plants are designed to encourage their activities. (For a description of bacteria that are less desirable in sewage-treatment plants, see chapter 11 on actinomycetes and chapter 7 on *Sphaerotilus*.) Most of these organisms are also found abundantly in other nutrient-rich environments.

Zoogloea, an alpha proteobacterium, is one of the most visible members of a healthy sewage-consuming community. If you have an opportunity to tour a treatment plant, you may be able to walk across a catwalk and look down on a large pool of agitated, aerated sewage covered by a thick blanket or floating rafts of foamy floc; *Zoogloea* is primarily responsible for that blanket of scum (plate 19). *Zoogloea* means "animal glue," which refers to the thick gelatinous matrix in which the bacterial cells are embedded. Sometimes the gel forms short branching fingerlike projections. Sewage in a treatment plant is an aerobic community that requires oxygen to be mixed in by agitation. In contrast, the communities that develop in the confines of home septic systems are anaerobic and are dominated by methanogens (chapter 2), although many other bacteria are present as well.

In addition to its role as a sewage organism, *Zoogloea*, along with other slime- and gel-forming bacteria (including *Alcaligenes*, a beta proteobacterium), are of interest as sources of biodegradable plastic.

VIEWING SEWAGE CONSUMERS UNDER A MICROSCOPE

Possibly, the sewage treatment operator will dip up a small sample for you. You should see a huge number of microorganisms, including stalked ciliates and worms. *Zoogloea* may or may not be identifiable, depending on whether it is making short gelatinous fingers. You may also see long, thin, threadlike filaments of *Sphaerotilus* (a beta proteobacterium) alongside thicker filaments of fungi, about ten times the width of those of *Sphaerotilus*.

PURPLE NONSULFURS

The purple nonsulfurs are alpha proteobacteria, with the exception of *Rhodocyclus* species, which are beta proteobacteria. Both alpha and beta types can be found in the same locations—anaerobic environments that are within the photic zone, including fresh, marine, hypersaline, or thermal waters and sediments, especially those that are stagnant or eutrophic. Stratified waters or sediments are best for observing these bacteria, although they bloom (reproduce abundantly) only rarely and are often accompanied (and obscured) by purple sulfur bacteria.

The photosynthesis of purple nonsulfurs is somewhat unusual. Cyanobacteria (and algae and plants) use water as a source of hydrogen for converting carbon dioxide (CO_2) into sugar ($C_6H_{12}O_6$). Purple *sulfur* bacteria (chapter 9) use hydrogen sulfide (H_2S) as a source of hydrogen. Purple *non*sulfurs usually use organic molecules (essentially small food molecules) as their source of hydrogen.

If you see a purple bloom of bacteria and you smell hydrogen sulfide, that bloom is likely to be purple sulfur bacteria, though there may be purple nonsulfurs there as well.

Some purple nonsulfurs are marketed as cleaners of unpleasant-smelling wastes, such as those in lagoons for animal manure, eutrophic ponds, and sludge and leach fields. Purple nonsulfurs readily use some of the odorous organics in these places.

VIEWING PURPLE NONSULFURS UNDER A MICROSCOPE

Examine material from a pink bloom. You will most likely see lots of purple sulfurs, which can be identified by the white-yellow sulfur granules deposited inside (see chapter 9). If you happen to see spiral shaped bacteria *without* these granules, these may be purple nonsulfurs.

CULTURING PURPLE NONSULFURS

Purple nonsulfurs may be purchased under various brand names as purifiers of odiferous waters. However, a description of the bacteria rarely appears in the advertisement or on the label.

OTHER PINK-RED PHOTOSYNTHESIZERS

There are several groups of pink-red photosynthetic bacteria in the alpha proteos other than the purple nonsulfurs just described. Most do not have easily distinguishable field marks even though they may be present in great numbers in some environments. A good example is a group that includes *Roseobacter*, a pink photosynthetic alpha that is abundant on nearly every surface in coastal marine photic zones. *Roseobacter* is a major member of the "fouling" community that covers boat bottoms and wharfs as well as seaweed, sea grasses, and sediments. Any sterile (or at least newly scraped and painted) surface such as a boat bottom is quickly colonized by a film of attached organisms. So why are surfaces so seldom pink, if *Roseobacter* is a prominent fouling organism? Probably because other photosynthesizers are quick to attach and display their pigments; well-established fouling communities can also include large seaweeds and animals. Also, *Roseobacter* is only sometimes a photosynthesizer. In the presence of oxygen, it is a heterotroph and does not display its pigments prominently.

Some *Roseobacter* species cause gall-like swellings (similar to agrobacterial galls) on red seaweeds of the genus *Prionitis*, which are found on the West Coast, sometimes in kelp beds. It is not clear whether these galls cause any benefit or harm. *Roseobacter* seems to have had a long coevolution with its seaweed host, implying the possibility of a symbiosis. (*Prionitis* is one of the culinary seaweeds, so for humans the unsightly growth of galls due to *Roseobacter* is considered unfortunate.)

FOOD ENHANCERS

Most of the important food-enhancing bacteria are gram-positives, described in chapter 10. However, *Acetobacter* and *Gluconobacter* are our providers of vinegar. *Zymomonas*, another genus of alpha pro-teobacteria, is a fermenter sometimes found in naturally fermented tropical beverages such as palm wine and pulque (made with agave juice).

Vinegar (acetic acid) is the field mark of *Acetobacter*. By itself, vinegar is a useful food enhancer, although *Acetobacter* and its acidic product can spoil certain foods, such as wines and fruit juices. In their desirable form, acetic acid bacteria are found with yeast in kombucha, a fermented tea, and with fungi in palm wine (fer-mented from palm sap). The complex microbial brew known as Belgian lambic (described in chapter 10) also includes acetobacteria in its flavorful, aromatic mix of bacteria.

Although present-day Americans may dislike vinegar tastes in beverages, there are a great many traditional drinks of developing countries and of past cultures that are based on the acidity of vinegar. Even diluted vinegar itself has been used as a beverage. Perhaps this is not surprising, since lemonade and carbonated sodas are also highly acidic. One example is a beverage that used to be drunk during haying season in New England—a sort of poor man's lemonade called switchel, made from molasses, water, ginger, and vinegar. If homemade vinegar were used, there would certainly be a community of bacteria enhancing the mixture.

Nata de coco, a gelatin-like dessert, is made from sweetened coconut water with the help of *Acetobacter*. The jelling of the juice is a result of bacterial activity.

A non-food-related use of acetic acid bacteria is in the traditional method for tanning hides. These bacteria have a high tolerance for tannic acid (an extract of oak bark) and contribute their own acidity to the tanning process. Traditional tanning methods use a variety of concoctions including dog and fowl manure and decaying animal organs to attract and maintain a diverse community of microbes (in addition to acetic acid bacteria) that cure leather.

VIEWING FOOD-ENHANCING ALPHAS UNDER A MICROSCOPE

If you have some homemade vinegar or spoiled wine or cider, you might see some of these organisms. Also, when rotting fruit smells vinegary, it is due to the activities of acetic acid bacteria. Try looking for a white-brown film on the surface of the vinegar; this is likely to be a gelatinous mat of acetic acid bacteria. However, there may be many other bacteria as well, making it difficult to determine which gram-negative rod you are looking at.

CULTURING FOOD-ENHANCING ALPHAS

Try making your own vinegar. Sometimes a spoiled bottle of wine will undergo the conversion, becoming pretty good wine vinegar, or you can use homemade vinegar to deliberately introduce the right bacteria into wine or cider. One procedure for making vinegar involves floating a quarter slice of white bread on the surface of a pint of red wine and leaving it loosely covered with cheesecloth for about a month. White scum and a vinegary smell indicate the presence of acetic acid bacteria and a successful transformation of wine to vinegar. The scum, composed of bacteria, is called "mother" or "vinegar mother" and may be used repeatedly with fresh wine to produce more vinegar. Note that commercially produced vinegar is sterile and therefore not useful for culturing or microscopy.

MAGNETOTACTIC ALPHAS

The alpha proteobacterial genus *Aquaspirillum* and its relatives, inhabitants of stagnant fresh or marine waters and sediments, form crystals of magnetite (iron oxide; Fe_3O_4) inside their cells in the presence of reduced or dissolved (ferrous) iron. The magnetite consists of several particles, each enclosed by a membrane, which are arranged in a row and serve as magnets. *Aquaspirillum* apparently directs its movement using these magnets.

But where does *Aquaspirillum* have to travel that requires orientation by magnets? One hypothesis is that these magnetotactic bacteria position themselves in sediments, determining what is up and what is down by sensing north and south of the Earth's magnetic field. Having magnets seems to limit their swimming to two dimensions, thus enabling them to more quickly locate an optimum depth in water or sediments. They are microaerophiles, preferring low concentrations of oxygen. Almost everywhere on a magnetized sphere (such as Earth), the magnetic forces are indicators not only of magnetic north and south (which is not the same as geographic north and south) but of up and down. For bacteria in the Northern Hemisphere, north is a magnetic field that curves down, while in the Southern Hemisphere, south points down. At the equator the field is theoretically parallel to the Earth but in reality varies such that either north or south points down, depending on location. Magnetotactic bacteria were first discovered in the Northern Hemisphere and tended to head toward magnetic north when stimulated with a handheld magnet. Researchers then traveled to the equator and to Australia to see what magnetotactics in those areas were doing. True to the predictions, bacteria in Australia oriented toward magnetic south. At the equator there were magnetotactics with orientations both north and south, perhaps indicating that both types can orient either up or down, depending on local conditions.

CULTURING MAGNETOTACTIC ALPHAS

Look for magnetotactic bacteria in stagnant waters that contain anaerobic sediments. Duckweed floating on top is sometimes a promising sign. Take a sample of stagnant water and sediment in a jar such that sediment fills about one quarter of the jar and water the rest. Add a little duckweed on top if possible. Tape a bar magnet to the jar about halfway between the surfaces of the water and the sediment so that the north end is in contact with the glass (if you live in the Northern Hemisphere; otherwise place the south end in contact with the glass) (fig. 6.5). Cover the jar with foil. Wait several days to weeks, and then untape the magnet to see whether a tiny grayish spot of bacteria has appeared. If so, these are magnetotactic bacteria (see the next section for hints on microscopic viewing). Even if there is no spot, try taking a sample from the area near the magnet. Sometimes there is just a wispy, diffuse area on the glass. You can also try positioning the magnet higher or lower on the jar. Magnetotactics favor a zone between oxidized and reduced sediments or waters.

Figure 6.5. Tape magnets to a jar to enrich for magnetotactic bacteria.

VIEWING MAGNETOTACTIC ALPHAS
UNDER A MICROSCOPE

If you were successful in getting magnetotactic bacteria to accumulate at your magnet, take some of the material from the spot, place it on a slide with a coverslip, and look at it under high power. If you place your bar magnet to the side of the slide, you should notice that some of the bacteria travel quickly either toward the magnet's north pole or away from its south pole. You can cause the tiny, helical bacteria to switch direction by flipping the magnet back and forth.

**SUMMARY: FIELD MARKS AND HABITATS
OF ALPHA PROTEOBACTERIA**

- ourselves and all eukaryotes bearing mitochondria (former alpha proteobacteria)
- some galls and burls on herbaceous plants, shrubs, and trees (*Agrobacterium*)
- root nodules of legumes (*Rhizobium*)
- some insects with skewed sex ratios favoring females (*Wolbachia*)
- pinkish slimy or ropy masses in methane-rich areas (methanotrophs; but note that many different bacteria produce slimes)
- floating blanket of scum in a sewage treatment plant (*Zoogloea*)
- purple bloom in stagnant waters (purple nonsulfurs; may be obscured by purple sulfurs)
- purple bloom in a waste lagoon (purple nonsulfurs; beta proteobacteria may also be present)
- fouling film (if pink) on boat bottoms, wharfs, and seaweed (*Roseobacter*) (often obscured by larger fouling organisms)
- gall-like swellings of some red algae (*Roseobacter*)
- some foods and drinks, such as vinegar and Belgian lambic
- stagnant waters and anaerobic sediments, often with duckweed present (magnetotactics)

CHAPTER 7

Beta Proteobacteria

The "betas" are there, but you can't see most of them. Some are prodigious oxidizers of manganese and iron and become encrusted with these metals, thus revealing themselves. Others manifest themselves by corroding stone monuments and stone buildings with their acid wastes. That same corrosive group also turns out to be essential for the well-being of fish aquaria. Other than these "visible" species, there are many hidden or obscured betas in soil and water, as well as a few pathogens—including *Neisseria*, the cause of gonorrhea.

BACTERIA AND METALS

All organisms need trace amounts of metals—such as iron, manganese, and copper—for the proper functioning of many of their chemical reactions. Bacteria have a variety of mechanisms for scavenging metals from their environments, such as by means of iron-binding proteins. In most cases, however, metals are not accumulated in visible or easily detectable amounts and therefore do not constitute good field marks. Furthermore, the accumulation of metals is probably universal among bacteria, because bacterial cell walls are negatively charged and naturally bind to positively charged ions of all sorts, including metals in solutions. In most cases, these metals are present only in trace amounts in cell walls. Some bacteria, however, especially the proteobacteria and cyanobacteria living in metal-rich environments, create conditions that affect metals in the surrounding environment in ways that can produce good field indicators. (See chapter 13 for a discussion of how the waste product of cyanobacteria—oxygen—interacts with metals.)

Many bacteria (not just beta proteobacteria) alter conditions in their environments such that metals are put into solution or precipitated. This is usually a nonspecific process by which acidic byproducts of fermentation and anaerobic conditions help dissolve (or weather or leach) metals from rocks or metal pipes. Large amounts of iron dissolved in water may be detectable when the water contacts air, forming iron oxides (rusty discolorations) as a result. The water may appear reddish as well.

Beta proteobacteria produce the most recognizable and specific field marks that result from metal oxidation. Although other bacteria may be involved, such activities will be considered field marks primarily of betas. Some betas and other bacteria oxidize iron and other compounds, using the energy to synthesize food in an autotrophic metabolism. These bacteria are called chemolithoautotrophs and are described at the end of this chapter and in chapter 9. The most visible of the beta chemolithoautotrophs oxidize iron and ammonia.

METAL-OXIDIZING BETA PROTEOBACTERIA

With about 20% oxygen in our atmosphere, the oxidation of metals happens easily with or without the participation of bacteria. Copper oxidizes to a green verdigris; iron oxidizes to red rust. In some environments not directly exposed to the atmosphere, the beta proteobacteria play an interesting role in metal cycles by causing oxidation that might not otherwise occur, sometimes producing large deposits of metal. Iron and manganese are the metals most likely to be affected by soil and water bacteria. In some mining operations, bacteria are used to oxidize (and thus accumulate) other metals, such as copper, uranium, and gold. In marine environments and all types of mineral springs (thermal or not, underwater or aboveground), iron-rich, anaerobic waters and sediments are susceptible to oxidation by bacteria when the water or sediments encounter the atmosphere.

SWAMPS AND BOGS
Some soil and aquatic habitats are rich in iron, especially if there are underlying or nearby deposits of iron-rich rocks. If those rocks are in

contact with anaerobic, acidic swamps or bogs, iron may be dissolved from these rocks and mobilized into the water system, sometimes traveling great distances. The iron is reduced from ferric iron (Fe^{++}) to ferrous iron (Fe^{+++}). Fermenting anaerobic, acid-producing bacteria of all sorts contribute to the conditions for metal dissolution. Wherever iron-rich waters contact the air, iron oxidizes and forms a rusty-red deposit. Thus, rivers or streams that drain iron-rich swamps or bogs are sometimes reddish-brown to black, and the rocks exposed to oxygen have red deposits (plate 20). It is in the wet soil and water of these environments that field marks of iron-oxidizing bacteria can be observed.

Leptothrix, Sphaerotilus, and related genera of long, filamentous, sheathed bacteria and *Gallionella* (stalked bacteria) are beta proteobacteria that favor oxic/anoxic interfaces of fresh, eutrophic, and marine waters. Such interfaces include slow-running acidic waters that drain swamps, bogs, or sewage treatment areas. The bacteria can form thick flocs (floating, slimy rafts); woolly, filamentous "carpets"; and long, slimy tassels ("acid streamers") attached to rocks or wood, which may be visible to the naked eye. Note that such systems may also be hospitable to floating and attached masses of slimy green algae. However, the flocs and tassels of *Leptothrix* and *Sphaerotilus* may be reddish-brown with encrusted iron that has been oxidized and accumulated by these bacteria.

Extensive deposits of iron oxides called bog iron, a low-grade iron ore that was once economically important especially in swampy, boggy areas of the eastern United States, can be considered field marks of beta proteobacteria. These deposits form over a period of years partly as a result of the action of iron-oxidizing bacteria. Bog iron is a hydrated (watery) form of iron oxide that is also called by the mineral names limonite or goethite. It is a yellowish-brown, porous, amorphous deposit that is sometimes present in such quantities in acidic ponds and streams draining bogs and swamps that it can be raked or fished up. Bog iron and iron-rich clay (ocher) are processed by being packed into furnaces with charcoal and oyster shells and heated to temperatures of 1,000°F or more. Impurities (known as "slag") are drained from the bottom, leaving behind a sponge-like "bloom" that can be processed (by hammering or reheating) further to remove more slag from the iron-rich bloom (plate 21).

Red clays from such areas are full of oxidized iron and are used for redware pottery, flower pots, and bricks as well as yellow-red pigments. These, along with objects made from bog iron, are the indirect products of iron-oxidizing bacteria.

To look at the iron cycle in such an iron-rich area, first examine the water draining from the bog or swamp. It should be reddish-brown to black as a result of organic acids flowing from the un-decayed or partly decayed plant materials that accumulate in acidic wetlands. The blackness results from the abundance of multicolored, undecayed material, all those colors together absorbing all of the wavelengths of light and yielding black. These acids dissolve iron, which is then oxidized wherever oxygen is present—near partly emerged rocks and plants or on shallow bottoms—into reddish iron oxides, which settle out with or without the help of organisms.

The state of iron compounds in soils and sediments can be determined by their colors and consistencies. Soils and clays that are yellow-orange to reddish-brown contain various forms of oxidized (ferric) iron or rust. These soils are sometimes called ochers and can be used as a low-grade source of bog iron. Blue-gray to greenish soils or clays are "gleyed" and are indicative of a relative lack of oxygen and the presence of dissolved (ferrous) iron (plate 22). Gleyed soils may show spots of red oxidation, especially in the vicinity of plant roots and sand or rocks, which make the soil more porous. Where oxidation is more extensive, bog iron may be deposited. Although iron oxidizes rather spontaneously in an oxygen-rich atmosphere, iron-oxidizing bacteria—especially those of the beta proteobacteria —enhance this process. These include the genera *Gallionella*, *Leptothrix*, and *Sphaerotilus*.

Some of these same betas oxidize manganese, a metal that behaves in a way similar to that of iron. One indication of oxidized manganese is a black precipitate on plant roots in an area of gleyed soil. A particularly visible example of manganese oxidation can sometimes be observed in small, still pools and ditches in the vicinity of iron bogs. Look for a shiny, silvery film on the water surface that has the appearance of an oil spill (plates 23, 24). To determine whether it contains oxidized manganese along with iron, simply touch it; it should break into jagged fragments. *Leptothrix* and *Siderocapsa* are sometimes the oxidizers responsible for iron-manganese films of this type. Communities of organisms in surface films such as these are called neuston communities. (See chapter 18 for additional information on interpreting these communities.)

Sphaerotilus in a Non-Iron-Rich Environment

Sphaerotilus was mentioned in the previous section as one of several iron-encrusted filamentous bacteria of swamps and bogs. However, becoming encrusted with iron is just one of the possibilities for *Sphaerotilus* species. These are heterotrophs that thrive on any abundant influx of rich food, such as sewage. In fact, they are sometimes incorrectly called "sewage fungus" after their tendency to form masses of slimy filaments in sludge. These masses cause a problem called "bulking," which prevents sludge from settling properly. Therefore, *Sphaerotilus* may be observable with or without iron deposits. If you have access to a stream of wastewater, such as from a paper mill or agricultural runoff, you may have an opportunity to see these bacteria. On a visit to a sewage treatment plant, *Sphaerotilus* may be pointed out to you by a knowledgeable guide (perhaps incorrectly referred to as a fungus) as well as *Zoogloea* (see chapter 6).

Plants as Indicators of Metal Oxidizers

Plants rely on trace amounts of metals in the soil to make some of their molecules, and deficiencies in these metals may result in discolored leaves. (Note that the diagnosis of metal deficiencies in plants is complex enough that this section cannot be a definitive guide.) Usually there is no direct connection between that phenomenon and specific bacterial activity; however, certain colors in leaves accompanied by particular soil conditions may be taken as evidence of bacteria that oxidize iron or manganese. These metals in their oxidized states are generally not available for uptake by plants. In their reduced (dissolved) state, these same metals may have leached from the soil, leaving behind metal-poor bluish clay, or gley. When you see plants near or in waterlogged, boggy, or swampy soil, especially plants that are grown for agricultural purposes, look for particular discolorations of the leaves, described below. One of the general terms for deficiencies of this sort, especially in peat soil, is "bog disease." Such a situation might occur in flooding conditions or at the edge of a field that borders closely on a marsh, swamp, or bog. Plants that are native to wetlands are likely to have mechanisms for dealing with deficiencies, and therefore are not good indicators of metal-oxidizing bacteria.

Iron deficiency (and therefore the activity of iron oxidizers) is suggested by a chlorosis (yellowing) of leaf veins as well as yellow or white spots on areas between the veins. Young leaves may be completely yellow or white, while older leaves, if they developed before the deficiency began, may still be green. The phenomenon is called "iron chlorosis."

A deficiency of manganese goes by colloquial names such as "marsh spot," "speckled yellows," and "gray speck." The leaves may be mottled gray or brown between the veins or yellow-white starting at the leaf margins. The veins themselves remain green. As with iron deficiency, leaves that developed before the deficiency took place may look normal. Note that any of these deficiencies can occur simultaneously, rendering diagnosis a challenge.

Other bacteria are associated with different metal deficiencies. For example, sulfur deficiencies sometimes occur in boggy areas where most of the sulfur is bound as iron sulfides. If you can smell hydrogen sulfide in an area with dark sediments delta proteobacteria (chapter 9) are probably present, and sulfur deficiency in nearby plants is a possibility. Look for a complete yellowing of leaves, including the veins.

Lack of bacteria can also cause problems for plants that are native to boggy areas but are transplanted to soils that are not acidic enough. For example, swamp and black tupelos become chlorotic (yellow) if their soil is too neutral in pH and lacks an assemblage of bog-type bacteria.

IRON SPRINGS AND SEEPS

Some mineral springs or seeps, ranging from cool to hot, are good sources of dissolved (reduced) iron and are good places to look for iron-oxidizing bacteria (plate 25). Iron seeps may also be found on the walls of some caves and mines. When looking on maps or in guidebooks for iron-bearing springs, look for adjectives such as "chalybeate" (iron-bearing), "ferric," "ferrous," or "iron." Sometimes so much iron is available that red-brown flocculent mats can be found floating on or thickly attached to surfaces. Because simple contact with oxygen-rich water, sediment, and atmosphere is enough to make iron rust and leave a conspicuous red deposit, not all the red iron at an iron spring can be attributed to bacterial oxidation. However, iron-oxidizing bacteria are very likely to be present, taking advantage of the interface between dissolved and precipitated iron. Iron springs may produce black manganese deposits as well, though these might be obscured if iron is the primary metal in the water.

MANGANESE NODULES IN AQUATIC ENVIRONMENTS

Lakes that are deep enough to contain stratified zones of oxidized and reduced (oxygen-poor) sediments or waters might have a manganese cycle. In the reduced zones, manganese goes into solution; in oxidized areas, it becomes manganese oxide and precipitates, often along with iron, which is cycled in the same way. Manganese-oxidizing bacteria sometimes form black lumps or nodules of oxidized iron and manganese—called "ferromanganese nodules"—which may be found in and on the lake sediments. The same sort of nodule formation may occur in the deep sea, though these areas are usually inaccessible except to research submarines. Even lake nodules are somewhat off limits to amateurs because they require positive identification of a mineral—manganese oxide. Tentative identifications may be made based on color and weight: look for nodules that are black to dark brown (both outside and within) and that weigh somewhat less than rocks of similar size.

WASTEWATER PIPES

Sometimes wastewater pipes become encrusted with oxidized iron and oxidized manganese deposits or tubercles. Such deposits are actually made of complex communities of bacteria—sometimes incorrectly referred to as "sewage fungus." Bacteria in these communities include not only metal oxidizers such as *Gallionella* (a beta proteobacterium) but also ammonia oxidizers and sulfate-reducing bacteria (delta proteobacteria; see chapter 9). However, the prominent field marks—partial or complete blockage of the pipe with encrustations, rusty slime, and rusty water—are likely to be those of beta proteobacteria. Any sulfur smell and black discoloration can be attributed to sulfate reducers. Note that "sewage fungus" is a colloquial term that is also used to describe an inappropriate bloom of *Sphaerotilus* in a sewage treatment plant.

MARINE ENVIRONMENTS

In marine environments, anaerobic sediments may be very rich in black iron sulfide (pyrite), which is oxidized to red rather quickly at exposed surfaces, with or without bacteria present.

A striking example of iron oxidation by marine bacteria may be seen (at least in photographs) on some wrecks of iron ships; in some cases, stalactite-shaped structures composed of iron oxides and bacteria hang from the corroding surfaces. Old piers often have iron supports or cables that are oxidizing with some assistance from bacteria (plate 26).

You may be able to find mollusc shells that are stained red-brown with iron oxide. In the mudflat gastropod *Hydrobla ulvae*, these deposits are known to be produced by bacteria. This phenomenon may be widespread, so it is reasonable to conclude that any reddish-stained molluscan or crustacean shell might have bacterial involvement (plate 27). Manganese oxide and iron oxide films may also form in marine settings. Bacterial oxidizers in such environments include iron- and manganese-oxidizing *Gallionella* and its relatives (plate 28). However, it can be difficult in a marine setting to distinguish spontaneous oxidation from that which is mediated by bacteria. Normal seawater is a rapid oxidizer of metals.

DESERT VARNISH

A black film coating the rocks or "pavements" (hard substrates) of deserts is sometimes the work of manganese oxidizers (plate 29). The coating, called "desert varnish" or "desert lacquer" is found in areas that receive intermittent water, such as stream beds. The varnish is a mix of manganese and iron oxides that are produced in part by metal-oxidizing bacteria. Other bacteria, including cyanobacteria, and lichens may also be involved in producing desert varnish. Note that any dry, brown-black crust on a desert rock probably consists of dormant cyanobacteria or lichens waiting for rain, and is not a mineral deposit. As with manganese nodules, an evaluation of the exact composition of desert varnish is somewhat beyond the purview of this book.

Chemolithoautotrophy

Chemolithoautotrophy is an extraordinary type of metabolism carried out by only a few groups of bacteria. "Chemo-" indicates that such organisms use chemicals rather than light as a source of energy. "Litho" (literally, rock) tells us that the specific chemicals are often minerals that may be found in some rocks and soils or else dissolved in water. "Autotrophy" means that these organisms manufacture their own food rather than seeking out and consuming food molecules from some external source. The multi-syllabic name serves as a mnemonic aid for this exotic (from our human-centered point of view) metabolism.

In table 7.1, chemolithoautotrophy (sometimes shortened to chemoautotrophy) is contrasted with the two major and more familiar forms of metabolism: photoautotrophy (photosynthesis) and heterotrophy. Table 7.2 focuses on the best bacterial field marks for chemolithoautotrophy. Note that the best field marks of chemoautotrophy are produced by the methanogens (chapter 2); by some of the beta proteobacteria, including ammonia and nitrite oxidizers and iron and sulfur oxidizers; and by sulfur-oxidizing gamma proteobacteria (chapter 9). Although chemolithoautotrophy is found in other bacterial groups, it is the focus in this guide only for those groups that produce macroscopic evidence of it. Chemolithoautotrophs that oxidize hydrogen are widespread in many groups but for the most part do not have sufficiently distinctive field marks for inclusion in this guide. An exception might be the knallgas alpha bacteria (see chapter 6), some of which produce slimes in methane-rich environments.

Chemolithoautotrophy is of great interest to planetary geologists seeking evidence of extraterrestrial life. Even on a dark, cold planet, such a metabolism might take place if there were a hot core and a supply of the right minerals. An extraterrestrial community containing chemolithoautotrophs as well as the heterotrophs that consume them might look something like the deep sea–vent communities described later in this chapter. Understanding the field marks of chemolithoautotrophic communities on Earth is of course a necessary first step to seeking them on other planets.

THIOBACILLUS AND ITS RELATIVES

Thiobacillus species are among the beta proteobacteria able to oxidize iron (or sulfide or copper) and use the chemical energy released to make sugar from carbon dioxide. *Thiobacillus* and other

TABLE 7.1. Types of Bacterial Metabolism

Source of energy	Metabolism type (source of food—carbon compounds)	Name for organism with this metabolism type*	Common name or shortened form	Who does it?
Chemical (chemo-), from food molecules (organo-)	Heterotrophy (consumes from environment)	Chemoorganoheterotroph	Heterotroph Consumer	Animals Fungi Many protists Many bacteria
Chemical (chemo-), from minerals (litho-)	Autotrophy (makes own)	Chemolithoautotroph	Chemoautotroph Lithotroph	Some bacteria
Light (photo-)	Autotrophy (makes own)	Photoautotroph**	Photosynthesizer Autotroph (usually used by nonmicrobiologists)	Plants Algae Many bacteria

* Some bacteria have more than one source of food. For example, halophiles (chapter 4) are photoheterotrophs because they consume food molecules but also collect light energy (although not for food synthesis). Some *Beggiatoa* are chemolithoheterotrophs because they consume food molecules but also break down minerals, perhaps deriving some useful energy, although not for food synthesis.
** Sometimes "litho" or "organo" is inserted into photoautotroph (as in photolithoautotroph) to distinguish two sources of carbon in the synthesis of food. Litho = carbon dioxide; organo = some small organic acid or other molecule.

TABLE 7.2. Chemolithoautotrophs: Energy Sources and Waste Products

Mineral or inorganic compound used as source of energy*	Waste product	Who does it?
H_2 (hydrogen)	CH_4** (methane)	Methanogens
H_2	H_2O	Hydrogen oxidizers (including alpha proteobacteria)
H_2	CH_3—COOH (acetic acid)	Some acetogenic bacteria (alpha proteobacteria***)
CO (carbon monoxide)	CO_2	Carbon monoxide oxidizers (gamma proteobacteria)
NH_4^+ (ammonium)	NO_2^-** (nitrite)	Ammonium oxidizers (beta proteobacteria)
NO_2^- (nitrite)	NO_3^-** (nitrate)	Nitrite oxidizers (beta proteobacteria)
Fe^{2+} (iron)	Fe^{3+}*** (rusted iron)	Iron oxidizers (beta proteobacteria, e.g., *Thiobacillus*)
S or $S_2O_3^{2-}$ (sulfur)	SO_4^{2-}** (sulfate)	Sulfur oxidizers (beta and gamma proteobacteria, e.g., *Thiobacillus*, *Thiomicrospira*)

* Compound is oxidized with oxygen (O_2) or some other compound. Source of carbon for all types of chemolithoautotrophy is carbon dioxide (CO_2).
** A good field mark.
*** The autotrophic producers of acetic acid are different from those that produce vinegar (see chapter 6).

such bacteria, including some of the iron-oxidizing bacteria mentioned in the previous section (e.g., *Gallionella*), are essentially doing what photosynthesizers do—photoautotrophy—but rather than using light as a source of energy, these bacteria use the energy from chemical bonds in minerals. *Thiobacillus* uses only this energy source, but the metal-oxidizing bacteria in the previous section typically supplement this kind of autotrophy with heterotrophy.

Thiobacillus gains more energy from sulfides compounds than from iron or copper. If sulfides are in the environment, these are the compounds of choice for *Thiobacillus* and most related genera. Hydrogen sulfide, a waste product of sulfate-reducing bacteria, is abundant in many anaerobic sediments and waters, especially marine. In many of these environments, *Thiobacillus* preferentially oxidizes hydrogen sulfide to sulfates.

If iron sulfides, such as black pyrite (FeS_2), are available in anaerobic marine sediments, *Thiobacillus* may oxidize them, producing mainly iron sulfates ($Fe_2(SO_4)_3$). In fact, some species of *Thiobacillus* may use metal sulfides exclusively. Copper sulfide or copper selenide can also serve as substrates for some *Thiobacillus* species, but these metals are relatively scarce in sediments except in the vicinity of copper mining operations.

Look for *Thiobacillus* and its relatives in anaerobic environments that are sulfide or iron rich, in the interface between anaerobic and aerobic zones. *Thiobacillus* can thrive in very acidic waters, unlike most of the iron-oxidizing bacteria described previously. However, metal oxidation is not an easy source of energy for *Thiobacillus*, so these bacteria are not often abundant enough to display field marks. Photosynthesizers can accumulate enormous (often highly visible) blooms in such environments, obscuring other bacteria. Chemoautotrophs are much less likely to produce vast blooms; an exception might be at mines that use bacteria to extract copper.

Mining Operations
Some mining processes by which metals are extracted from ore take advantage of an ancient practice using the activities of metal and sulfide oxidizers of the *Thiobacillus* group. The archaean *Sulfolobus* is also sometimes used. Copper sulfide ores may be processed by having *Thiobacillus* oxidize the sulfide and iron to release the copper. The process, called chalcopyrite ($CuFeS_2$) leaching, is the source of 25% of the copper extracted from low-grade ore in the United States. Uranium is also sometimes processed this way, and the method is considered feasible for some ores of silver, gold, zinc, and nickel. Cost-effectiveness is always a criterion in using such techniques, though, and cheaper chemical extractions are sometimes used, even though the bacterial methods are often cleaner for the environment.

Methods for bacterial leaching include "heap leaching," in which *Thiobacillus* is circulated over a pile or "dump" of ore, and in situ leaching, in which *Thiobacillus* is pumped directly into a mine opening. Copper sulfate in solution is collected and converted chemically to metallic copper.

Sometimes another bacterial method is used to collect copper or other metals from leach water. Beta proteobacteria such as *Sphaerotilus* may be used to absorb copper in their sheaths (thick outer coverings) in a technique known as biosorption. In fact, large numbers of almost any kind of microorganism work well for the biosorption of metals, and cyanobacteria and eukaryotic algae have also been used with success. There is a natural tendency for bacterial walls and sheaths to bind metallic ions nonspecifically and without any known metabolic function. Biosorption is also used to remove heavy metals from wastewater, often by running the waste through algal or cyanobacterial mats (chapter 13).

Ancient Ore Deposits

Many ancient deposits of metal ores are well understood by geologists to have arisen by some geochemical or geophysical process. Some deposits, however, cannot be explained entirely except by invoking the activities of ancient bacterial communities. Massive iron deposits known as "banded iron formations" (BIFs) and "red beds," for example, can be attributed to the production of oxygen by cyanobacteria. Most of the world's iron is extracted from these deposits (chapter 13).

Some deposits of uranium, copper, and gold may have bacterial origins as well. Gold is used here as the example. About $2^1/_2$ billion years ago, large amounts of gold were deposited in river deltas in a way that has not been seen since. These ancient deposits are rather fine and form a thin layer that is in close association with microfossils of bacteria. This layer is not like the sort of particulate ("placer") gold that sometimes travels in modern rivers. One of the largest of these deposits is in South Africa. It appears that this gold may have been dissolved by bacteria of the *Thiobacillus* group. The gold then traveled in solution downstream, where it was trapped in the cell walls and sheaths of cyanobacteria living in mats in the river delta. This process mirrors that used in microbiological mining operations described previously. That is, *Thiobacillus* leaches the gold into solution from sulfide-rich rocks, and mats of cyanobacteria absorb it into their sheaths.

Oil Fields

Thiobacillus is sometimes used to process or "rejuvenate" oil shales to release more oil; it makes the rock more porous by dissolving metals. Various other sulfide oxidizers (chapter 9) can then be used to remove reactive sulfides from the crude oil.

The remediation of oil spills can also be a bacterial process, although a surprisingly nonspecific one. Many different bacteria can break down petroleum products, and these bacteria are readily available in most environments where oil spills occur. In fact, the ability to break down these products is such a general trait that entrepreneurs have found it difficult to make a profit selling special bacteria to clean spills. Fertilizing waters or soils is sometimes enough to stimulate bacteria to break down the chemicals in a spill, by providing nutrients not available in petroleum compounds. In a sense, this capability of bacteria is not surprising, given that petroleum products are simply the remains of plants that would have been broken down by ancient bacteria had they not become so deeply buried; modern petroleum-eating bacteria are simply completing a loop begun by their ancient counterparts. Indeed some petroleum-consuming bacteria are considered to be undesirable "pests" in oil fields and refineries.

Corrosion of Concrete and Marble

Some *Thiobacillus* species are responsible for the corrosion of concrete sewage systems. If hydrogen sulfide is present, *Thiobacillus* and other bacteria convert it to sulfur and, if conditions are acidic enough (which they are likely to be in sewage), to corrosive sulfuric acid. Thus, deteriorating concrete sewage systems are field marks of beta proteobacteria.

Although sulfates are not a typical component of marble, sulfur dioxide may be available in a polluted urban atmosphere. Under these conditions, *Thiobacillus* can convert sulfur dioxide to sulfuric acid, which dissolves marble (calcium carbonate) into soft, crumbly calcium sulfate. The result is a disaster for marble statuary and architectural features. (For descriptions of other attacks on art objects see this chapter on ammonia-oxidizing bacteria, chapter 11 on actinomycetes, and chapter 13 on cyanobacteria. Fungi can also cause serious damage to artwork, either on their own or in concert with various bacteria.)

CULTURING METAL-OXIDIZING BETAS AND *THIOBACILLUS*

Try a modified Winogradsky technique (as described in appendix A). In a jar, collect some promising waters and sediments from the vicinity of a bog or iron spring or from water flowing through iron pipes. Add straw to form a loose matrix of criss-crossed stems from the surface of the water to the sediment (plates 30, 31). You may need to add a source of iron or manganese, depending on the amount in the water and sediments. You can do this by burying an iron object in the sediment; mixing in small amounts of iron salts, such as $FeCl_3$, or manganese salts, such as $MnSO_4$ (available from a chemistry lab); or adding vitamin tablets rich in iron. (To enrich for sulfate-reducing bacteria as well, add a few Epsom salts, a source of magnesium [not manganese] sulfate. These are available at pharmacies.)

A more typical Winogradsky column of the sort intended to encourage colorful layers of sulfur bacteria (see chapter 9) may also display iron oxidizers. To encourage them, add a relatively small iron source, such as an iron nail stuck down through the sediments of the column, and look for a layer of oxidized iron. This usually forms between the top layer of cyanobacteria and the second layer of purple sulfur bacteria.

In the field, you can try burying iron objects such as pipes and railroad spikes in bogs or swamps or in iron-rich waters to see the effects (plate 32). Pull the objects up days or months later and look for films and crusts of iron oxide or corrosion. If you happen to find iron objects discarded in a swamp, dig them up and examine them. Any fountain or other architectural feature that is made of rusting iron and receives a trickle or spray of water might support bacteria. If sulfide is present in the environment, the object may give off a strong smell of hydrogen sulfide—the product of sulfate-reducing bacteria living under a crust of iron oxide and iron-oxidizing bacteria. Black iron sulfide may also be present, especially in marine waters.

You may also suspend in water all sorts of nonmetal objects. In a good iron bog, such items should get a nice coating of bacteria, including some that leave red-brown iron oxide or black manganese oxide deposits. If you suspend glass microscope slides in water, iron-oxidizing bacteria may attach and produce oxidized metal deposits. In almost any soil or water environment, you can improvise with soil test kits (sold at garden centers) to test for pH, nitrate, phosphate, and potassium (plate 33). For example, in an acidic iron-rich bog, expect low pH and low nitrates and potassium. In general, bogs have slow nutrient cycles, and nitrogen compounds are usually quite limited. However, phosphate may be high because it tends to bind iron as ferric phosphate, detectable by soil phosphate tests.

VIEWING METAL-OXIDIZING BETAS AND *THIOBACILLUS* UNDER A MICROSCOPE

Prepare some glass slides by suspending them in iron-rich waters (as just described). Wipe one side clean, and put a coverslip on the other side of the slide, which you should keep moist. Alternatively, you may take bits of sediment, scrapings from rocks, metallic surface film, or rusty-looking film from plant surfaces to examine under the microscope.

You should be able to see a diverse community of microorganisms. This type of communty is sometimes referred to as an "aufwuch" (attached) community, or, if it is in a surface film, a neuston community (see also chapter 18). Your slide or scraping may be full of eukaryotic microbes too, such as rotifers and stalked ciliates. If the slide or other object was in a photic zone, there may be prokaryotic and eukaryotic photosynthesizers. To see the iron- or manganese-oxidizing bacteria, look for stalked or filamentous bacteria, possibly encrusted with yellow to red-brown metal. Both *Sphaerotilus* and *Leptothrix* make long multicellular filaments and are enclosed by sheaths in which iron oxides may be found, although metals will not necessarily be identifiable with ordinary microscope techniques. *Gallionella* has a long, twisted stalk, a modified cell, at the end of which is a small bean-shaped cell; *Siderocapsa* is a small single cell. Both can accumulate iron oxides, and *Gallionella*, by virtue of its stalk, is reasonably identifiable.

However, see chapter 18 on the other stalked bacteria and eukaryotes that attach to surfaces.

AMMONIA-OXIDIZING CHEMOLITHOAUTOTROPHIC BETAS

Beta proteobacteria such as *Nitrosomonas* and *Nitrobacter* participate in the part of the nitrogen cycle in which ammonia is oxidized to nitrite (by *Nitrosomonas* and its relatives) and then to nitrate (by *Nitrobacter* and its relatives). Ammonia oxidizers are ubiquitous in any aquatic or soil environment where organic materials are being broken down. Sometimes an especially rich environment, such as a chicken house full of feces, smells of ammonia; ammonia oxidizers are sure to be present. Ammonia is available as the product of many organisms. Heterotrophs of all kinds (including animals) produce ammonia compounds as waste products of metabolism. Thus, almost any nutrient-rich environment is likely to be a source of ammonia. Nitrogen fixers (alpha proteobacteria) may also contribute to the cycle by converting nitrogen gas (N_2) to ammonia. If ammonia is produced in or diffuses into aerobic environments, *Nitrosomonas* and *Nitrobacter* species oxidize it and release enough chemical energy to produce food (sugar) from carbon dioxide. Thus, ammonia-oxidizing betas are chemoautotrophs—similar to photoautotrophs except that they use chemical energy instead of light energy.

Stone Monuments

A good field mark for ammonia oxidizers is an urban one—corroded stone monuments and buildings. These structures become corroded by the nitric acids produced by ammonia oxidizers. A layer of decaying lichens, algae, or bird guano provides a source of ammonia for the bacteria. Sometimes the rocks themselves are the source of ammonia, having been formed from partly decomposed microorganisms. Ammonia oxidizers that attack these rocks are essentially releasing that ammonia and helping to complete a part of the nitrogen cycle abandoned long ago.

Rounded edges, pitting, and powdering of what were once sharp corners and smooth surfaces are good indicators of ammonia oxidation. In particular, look for white, feltlike powders when dry, and a somewhat slimy texture on wetting. Pits may look reddish if iron oxide is present. This phenomenon may be considered one of the "diseases" of stone, or *Steinkrankheiten*, a term coined by German geomicrobiologist Wolfgang Krumbein. An old graveyard encrusted by lichens is a good place to look for pitted, powdered, and rounded stone carvings (plate 34). Another German microbiologist, Hans Schlagel, wrote that his boyhood interest in microbiology began when he observed powdery saltpeter (potassium nitrate) forming on a brick wall near a manure heap due to the activities of ammonia oxidizers.

Limestones and building materials composed of limestone (e.g., cement) are especially susceptible to degradation by nitric acid from ammonia oxidizers. However, there may be nonbiological sources of other acids. If sulfur-rich fossil fuels are being burned, sulfuric acid may precipitate on statues and buildings.

Fish Tanks

Healthy fish in your aquarium are a field mark of ammonia oxidizers. One of the challenges of keeping fish is controlling the accumulation of their waste products. Fish excrete ammonia from their gills into the aqueous environment. In natural bodies of water, wastes do not usually accumulate. In the confines of a tank, however, ammonia can build up to levels high enough to kill the inhabitants. Fortunately, ammonia oxidizers are nearly ubiquitous in soils and waters and are most likely present in any aquarium, even if no particular care has been taken to establish them there. Serious keepers of aquaria do make an effort to establish and maintain ammonia oxidizers, especially if they want to keep many fish and plants in the most healthy and natural environment that can be approximated in a small space. When an aquarium is first initiated, ammonia (detectable with a hobbyist test kit) can climb to dangerous levels. At this point, it is important to encourage the growth of ammonia oxidizers, either by "seeding" the water with a little gravel from a well-established tank or by adding freeze-dried bacterial cultures from an aquarium shop. A "biological" filtering system retains bacteria beneath the gravel and forms an aerated habitat in which the ammonia oxidizers can function efficiently.

After a couple of weeks, high levels of ammonia are typically replaced by high levels of nitrite, the product of *Nitrosomonas* and related oxidizers. The tank usually stabilizes with low levels of ammonia and nitrite and higher levels of (relatively nontoxic) nitrate produced by *Nitrobacter* and related oxidizers. The result is a well-established sequence of the nitrogen cycle: fish food to ammonia via the fish; ammonia to nitrite via *Nitrosomonas*; and nitrite to nitrate via *Nitrobacter*. All stages of this cycle should be monitored regularly by water testing. Whenever new fish are added, feeding regimes are changed, or antibiotics are used to treat infection, the bacterial community may fluctuate, altering levels of ammonia, nitrites, and nitrates.

VIEWING AMMONIA OXIDIZERS UNDER A MICROSCOPE

If bacteria on stone monuments and buildings are mixed in with bird guano or lichens, it will be difficult to distinguish them under the microscope from other types of bacteria.

There may be a great many different bacteria in the filter of your fish tank. It will not necessarily be obvious which are the ammonia oxidizers.

CULTURING AMMONIA OXIDIZERS

Maintaining a healthy aquarium with a well-established bacterial filter is essentially a culture method for ammonia oxidizers.

SUMMARY: FIELD MARKS AND HABITATS OF BETA PROTEOBACTERIA

(*Note*: In general, a reddish-orange color indicates iron-oxidizing bacteria, and black or silvery films or deposits indicate manganese-oxidizing bacteria, although both types are often present together.)

Acidic bogs, swamps, and boggy soil

- plants with yellow or speckled leaves, indicating iron or manganese deficiency
- rusty-red deposits (iron-oxide) in runoff or on rocks; bog iron
- floating, reddish-brown, slimy rafts or tassels (*Sphaerotilus*)
- shiny, silvery film on water that breaks into jagged fragments when touched

- black or red precipitate on plant roots in area of bluish clay soil (gley)

Mineral springs or seeps
- red iron deposits or red-brown flocculent mats
- black manganese deposits

Lakes (those deep enough to have stratified zones)
- brown-black lumps or nodules of manganese oxide in and around sediments

Marine environments
(Note that it may be difficult to distinguish bacterial action from spontaneous oxidation.)
- rusty, corroded structures on ships or piers
- red-brown stains on shells
- brown-black lumps or nodules of manganese

Other habitats
- black discoloration ("desert varnish") on desert rocks (manganese oxidizers)
- "sewage fungus" (slimy streamers) in sludge and other nutrient-rich waters (*Sphaerotilus*)
- encrustations or rusty slime in wastewater pipes
- deteriorating concrete sewage systems (*Thiobacillus*)
- biological mining processes and wastewater treatment processes (especially if heavy metals are being removed) (*Thiobacillus*)
- ancient ore deposits (product of *Thiobacillus* and other bacteria)
- oil-shale processing (*Thiobacillus*)
- crumbling marble or architectural features (*Thiobacillus*)
- corroded stone monuments, especially when pitted and powdery (ammonia oxidizers)
- healthy aquaria with "biological" filters (ammonia oxidizers)

CHAPTER 8

Gamma and Delta Proteobacteria of Non-Sulfur-Rich Environments

The gamma and delta proteobacteria that are not part of sulfur-rich communities (see chapter 9) produce intriguing field marks in a diversity of habitats—including inside insects and shipworms, and on human skin. These proteobacteria include three groups that, when described, seem like the invention of a science fiction writer: gammas that glow in the dark (bioluminescent gammas), gammas that influence the formation of snow and ice, and deltas that form multicellular treelike structures almost a millimeter high (gigantic on the bacterial scale).

MARINE BIOLUMINESCENT GAMMAS

Photobacter and bioluminescent *Vibrio* are genera of marine microbes that give off light, like tiny fireflies. All sorts of good experiments have demonstrated that fireflies use bioluminescent signals for mating purposes. However, such explanations, involving communication of some sort, don't work as well for individual bacterial cells. In a sparsely populated medium such as marine water, where most bacterial luminescence occurs, the light of one bacterium would be undetectable to other bacteria or to any other organism. Moreover, bacteria simply do not swim long distances (many bacterial body lengths) to get to "mates." Also, in the dilute water of the ocean, nutrients may often be too scarce to allow bacterial bioluminescence—producing light is a fairly expensive chemical reaction. "Why do it?" is a question that has entertained many microbiologists since these bacteria were discovered.

The first part of the answer to this question is bound to be a little disappointing, but don't worry—it gets much better. It turns out that the light-producing chemical reaction is a relatively simple extension of the metabolic reactions performed by all aerobic cells. A minor change in these reactions that would result in the production of photons (light) is really not so extraordinary, and indeed all chemical reactions give off small numbers of photons; bioluminescent organisms just happen to do so on a perceptible level. The fact that this sort of reaction has apparently evolved more than once, and with significantly different chemicals, in bacteria, some fungi, fireflies, dinoflagellates, and some marine animals suggests that it happens relatively easily and that, in the case of the bacteria, the light is merely serendipitous.

However, serendipity explains only how bioluminescence might have evolved as a side effect. For the phenomenon to have persisted, it would have to have some benefit for the organism. The reaction takes energy and oxygen, and thus could be considered a waste of resources unless put to some good use.

Detoxifying oxygen may be one of the primary functions of bioluminescence in bacteria. Oxygen is detrimental to cells except during its brief role in respiration, the set of reactions by which aerobic heterotrophs (such as ourselves) use food. Oxygen literally oxidizes and destroys the delicate chemicals of organisms; many specialized reactions in the cell are detoxification reactions that use up oxygen, thereby preventing it from doing damage. Bioluminescence may be one such reaction.

There may be another function of bioluminescence, though. Some sort of communication may in fact be going on, but not from bacterium to bacterium, and not in the open water. Like many aerobic heterotrophs, bioluminescent bacteria are opportunists; they flourish if they happen upon a rich source of food, such as a decaying fish. If great numbers of bacteria accumulate, the fish will actually glow with feeding microbes. The glowing fish may well attract other fish that, by ingesting it, establish the bioluminescent microbes in their own digestive tracts. Indeed, the digestive systems of some marine animals are one of the places where bioluminescent bacteria are likely to be found. These bacteria are really just a type of enteric (digestive system) bacteria with a fancy metabolism.

The bacterial glow might function to attract carrion eaters. But why would the bacteria "want" to be ingested? Other than decaying organic material, the world is relatively devoid of freely available nourishment, with one important exception: the digestive systems of animals. These environments provide wonderfully rich, stable supplies of nutrients and as such are a favorite habitat of many opportunistic heterotrophs. It is easy to envision a cycle—advantageous for the microbe—by which a bacterium recruits new hosts by causing a former, dead host to glow as it decays. Thus, some bacteria may never be exposed to the dilute ocean waters. There is also evidence that the bacteria stimulate each other such that larger populations have a higher output of light per individual cell—a phenomenon called "quorum sensing."

These bioluminescent bacteria, once removed from the host's body via the feces, may continue to glow, although the light is not typically observable with the naked eye. Fecal material and dead plankton are constantly drifting down in marine waters. If the particles are visible, they are referred to as "marine snow." Divers, especially in tropical waters or in kelp beds, are familiar with this "precipitation," which can even form billows and drifts on the ocean floor. "River snow" and "lake snow" are analogous to marine snow but consist of particles that are usually not large enough to see with the naked eye. One might even extend the metaphor to "aerial" snow in the form of dust motes, often visible in sunbeams. Dormant bacterial and fungal spores are often found as "riders" on marine snow, with each particle forming a microbial community inhabited by bacteria and protists. Bacteria may help keep the particles cohesive by secreting sticky polymers. Using rain gauge–like devices, oceanographers capture marine snow and tinier particles for analysis. Some "snow" collected in this way glows with light detectable by sensitive photometers, although not by the eye. Presumably, at least some of that glow is produced by bacteria that once lived in an animal's intestines.

The final part of the story is best of all. Large concentrations of bioluminescent bacteria situated in specific sections of the host's digestive system can cause that organ to glow. If the host—a fish, squid, or tunicate—is at all transparent, that light can be used by the host for activities such as mate selection and hunting prey. In this case, there is a symbiosis between marine animals and bacteria: the bacteria get comfortable culture chambers, and the host acquires a new behavioral signal—which it can elaborate on with special lenses and shutters of tissue and culture chambers of modified organs.

Many of these hosts are deep-sea organisms, unlikely to be observed by naturalists lacking special diving equipment or research submarines. If you are interested in viewing these bacteria, phone your local aquarium and ask if they have any of the species listed in table 8.1. Alternatively, look for a glowing fish on the seashore at night or a bait bucket of glowing shrimp that are on their way past their prime: you will know that you are seeing bioluminescent bac-

---

TABLE 8.1. Taxa of Marine Organisms with Species That Have Symbiotic Bioluminescent Bacteria

---

Fish
    Macrouridae (grenadiers or ratfish)
    Moridae (morid cods)
    Merlucciidae (merlucciid hakes)
    Trachichthyidae (slime heads)
    Anomalopidae (flashlight fish)*
        *Anomalops**
        *Kryptophanaron**
        *Photoblepheron**
    Monocentidae (pinecone fish)
    Leignathidae (pony fish)
    Apogoniidae (cardinal fish)
    Ceratioidei (11 families of anglerfish)

Cephalopod molluscs
    Sepiolidae*
        *Euprymna**
    Loliginidae (luminescent ink)

Urochordates
    Pyrosomatidae
        Pyrosoma

---

* Most likely to be viewed at an aquarium. Others might be seen in dioramas.

Note: Many deep sea animals are self-bioluminescent and do not have symbionts.

Source: K. Nealson and J. W. Hastings, in Albert Balows et al., eds., *The Prokaryotes: A Handbook on the Biology of Bacteria* (New York: Springer-Verlag, 1992).

teria. (Beautiful luminescent waters sometimes visible at night in the wakes of boats or when swimming are due not to bacteria but to the wonderfully named dinoflagellate *Noctiluca milliaris* ["a thousand night lights"].) (An unusual case of bioluminescence is that of the deep-sea "dragon fish" *Malocosteus niger*. This species makes its own red light by which it sneaks up on prey unable to see red. To see red itself, the fish uses modified chlorophyll from *Chlorobium* [chapter 5], which becomes concentrated in the food chain.)

VIEWING MARINE BIOLUMINESCENT GAMMAS UNDER A MICROSCOPE

If you can get close enough to that dead fish or shrimp to get a little scraping, you will see many different bacteria! After all, this dead animal is a rich source of nutrients. Some of the *Vibrio* bacteria may be bioluminescent, but it will be difficult to distinguish these from all the other opportunists in your sample.

CULTURING MARINE BIOLUMINESCENT GAMMAS

Again, as you are approaching that same dead fish or shrimp, have in hand a cotton swab and a plate of moderately nutritious medium (perhaps using a little fish broth as a nutrient; see appendix A). There is just one problem: you will get such a diversity of bacteria on your plate—many of them not bioluminescent—that the whole malodorous exercise may not be worth it. If you would like to pursue this, look in the literature for isolation and culture techniques of bioluminescent bacteria, and give it a try. A pitch-dark room, with your eyes adjusted to darkness, is best for viewing luminescence on a petri dish. Freshly inoculated plates with new colonies tend to glow best.

Alternatively, you may try culturing bioluminescent bacteria on your own dead fish or shrimp. Buy or capture a whole fish (e.g., flounder) or shrimp, place it in a shallow dish, and cover it halfway with salt water, preferably from the ocean or a marine tank. Put the dish in a refrigerator for several days. If you must make your own salt water, the proportions are about 30 grams of salt to 1 liter of water, although actual seawater is more likely to yield results. Take the dish into a pitch-dark place, and, after patiently waiting for your eyes to adjust, try to detect a fish- or shrimp-shaped glow. If you see any such glow, you have bioluminescent bacteria.

Before refrigeration, when seafood might be kept a few days in a chilly basement, this phenomenon of glowing decay was observed and noted. Charles Dickens, in *A Christmas Carol* (1852), likens Marley's face in the knocker of Scrooge's door to a glowing lobster: "Marley's face . . . had a dismal light about it, like a bad lobster in a dark cellar." How many nonmicrobiologists have passed over that line, unable to decipher what image Dickens had in mind?

TERRESTRIAL BIOLUMINESCENT GAMMAS

To see symbiosis in action between bioluminescent bacteria and a terrestrial organism, you must happen upon a dead caterpillar, glowing softly in the dark and strangely unputrifying. (If you are walking around in the dark looking for faint biological light, you may be fortunate enough to encounter bioluminescent fungi as well.) The dead, glowing caterpillar was killed by a soil nematode (roundworm) of the family Heterorhabditae, which feeds on (and contains) the bioluminescent bacteria *Xenorhabdus* and *Photohabdus*. The worm invades caterpillars, thereby releasing the bioluminescent bacteria, which cause caterpillars to glow in death. The advantage to the nematode is that the bioluminescent bacteria limit competition for the dead body of the caterpillar by releasing chemicals that inhibit other bacteria and fungi. Once the caterpillar is dead and glowing, it lingers undecayed, providing a habitat for the nematodes that are feeding on the luminescent bacteria.

> **VIEWING TERRESTRIAL BIOLUMINESCENT**
> **GAMMAS UNDER A MICROSCOPE**

Try mincing up a bit of glowing dead caterpillar, if you can find one. Most likely any bacteria you see are bioluminescent gammas, because they will have inhibited the growth of others.

> **CULTURING TERRESTRIAL BIOLUMINESCENT GAMMAS**

If you want to see evidence of bioluminescent bacteria, try purchasing the nematodes, which are available as biological control agents against garden pests. Be aware that garden products are not always advertised as containing bacteria, however, so look for "Hb" or "Sc" on the label, indicating two common types of nematodes: *Heterorhabditis* and *Steinernema*. The word "nematode" is sometimes mentioned too, but not the fact that the nematodes make the corpses of caterpillars and grubs glow.

Gammas of Snow and Ice

Two genera of gamma proteobacteria, *Erwinia* and *Pseudomonas* (which also have members in the alpha and beta proteos due to lingering problems with their classifications), have been identified repeatedly as being capable of nucleating (forming the center for) crystals of ice at temperatures slightly warmer than the freezing point of water. The bacteria have unique proteins that seem to mimic the surface of an ice crystal, thus encouraging real ice to form there. These bacteria are prevalent on many plants and can damage frost-sensitive ones. They are also among the first bacteria to cause vegetables and other plant materials to spoil. Frost crystals break open plant cells and make them more accessible to bacteria. The bacteria may prevent themselves from freezing, however, by ensuring that the ice crystals are formed outside their own cells. Can you detect this phenomenon in your own garden? Perhaps, considering that these bacteria can be found on most plants that you might be growing. Therefore, the formation of frost in your garden can be considered a field mark of these bacteria—especially if the frost seems to be forming at a few degrees above the freezing point of water.

A controversial product called Frostban was designed to protect crops by removing the ice-nucleating genes of *Pseudomonas*. Another version of Frostban was composed of naturally occurring pseudomonads that happen to be missing the gene to nucleate ice. The idea for the Frostban project was to spray crops with the altered or deficient bacteria in the hopes that they would crowd out the normal population of pseudomonads, thereby giving the plants a few extra degrees of protection against frost. The product is still in a testing phase and has encountered some opposition due to the "modified" nature of the bacteria.

Snowfall might be considered another field mark of *Pseudomonas* and *Erwinia*. Dust in the atmosphere is known to nucleate or "seed" both raindrops and snow crystals. That is why seeding clouds from an airplane with silver iodide, dry ice, or tiny particles of paper pulp can cause or enhance precipitation. Typical atmospheric "dust" is quite heterogeneous, consisting of all sorts of nonbiological materials such as clay particles, as well as bacteria and fungal spores. Among the bacteria in this dust are *Pseudomonas* and *Erwinia*. These two genera have been demonstrated in laboratory settings to be effective seeders of snow, presumably because of their ice-nucleating proteins. Snow crystals may form more readily when these bacteria are present. In fact, the ice-nucleating protein has been isolated and sold as Snomax, which is sometimes added to snow-making machines at ski resorts and winter Olympic events. The bacterial protein helps produce lighter, dryer snowflakes.

VIEWING GAMMAS OF SNOW AND ICE UNDER A MICROSCOPE

It is difficult to identify these bacteria in plant material after a frost. These bacteria coexist with many other bacteria and fungi decaying the plant material. In snowfall, they are too dilute.

GAMMAS OF THE SKIN

Most skin bacteria are gram-positives (see chapter 12), but the gamma proteobacterium *Acinetobacter* is a notable exception. On the human body, *Acinetobacter* is most likely to be found in the armpits, groin, and feet—especially between the toes if they are damp enough. *Acinetobacter* is also found in soil and water. However, as one of many tiny rods in these moist places, these bacteria are difficult to distinguish from other more abundant and more odoriferous inhabitants.

GAMMAS AND SHIPWORMS

Shipworms (*Teredo*), which make the smooth honeycomb borings in wharves, wooden boats, and driftwood (fig. 8.1), are not worms at all but bivalve (two-shelled) mollusks of the same group as clams.

FIGURE **8.1.** *Teredo*, or shipworm, the creature responsible for these tunnels, is actually a bivalve mollusc. It has symbiotic gamma proteobacteria that help it digest wood.

Nevertheless, this bivalve has evolved a very wormlike habit. It is elongate and uses two greatly reduced shells to bore its way through wood—which is both its habitat and its source of food. Thus it joins the ranks of cellulose eaters and especially wood eaters in having a challenging source of nutrition. With the exception of fungi, eukaryotes do not completely digest cellulose, especially in the form of wood. Termites and various ruminating animals—and in fact nearly all herbivores—have the same difficulty with digesting cellulose, which explains why so many are built like fermentation vats full of cellulose-digesting microbes (e.g., *Bacteroides* and gram-positives; see chapters 12 and 15). In the case of shipworms, the symbionts are gamma proteobacteria (genus *Teredinobacter*) that both digest cellulose and fix nitrogen (convert gaseous nitrogen into a form usable for making protein). Because these bacteria are essential for the normal functioning and well-being of their hosts, you may consider wood riddled with shipworm borings to be a field mark of these gamma proteobacteria.

Gammas as Symbionts of Aphids

Aphids are interesting insects made even more so by their bacterial symbionts. Aphids are small, pale, soft-bodied insects that are often found in great numbers on plants. They bore into plants and then suck up plant juice through tiny probosces. Their reproduction is by parthenogenesis: populations are composed of only females, which are able to reproduce daughters without fertilization by process of "live birth" called vivipary. After many generations, in some species, males are formed and reproduce sexually with females. This switch in sexuality is often accompanied by a change to a different plant for food. Other aphid species are permanently parthenogenetic.

All aphids have as symbionts the highly modified gamma proteobacterium *Buchnera*. This bacterium apparently supplies the necessary nutrients to supplement the aphids' sugary diet of sweet, watery plant sap with little protein, fat, or complex carbohydrates. (Other sucking insects with restricted diets also have their own particular bacterial symbionts.) *Buchnera* is passed from mother to daughter, which may in part explain the evolution of parthenogenesis in this group: it ensures the efficient propagation of symbiotic bacteria. *Buchnera* "stuck" inside a male would essentially have reached a dead end, being unable to exit via the sperm.

In addition to symbiotic *Buchnera*, many aphids may also harbor *Wolbachia*, an alpha proteobacterium (chapter 6). These bacteria also have a "motivation" for preferring females because of their transmission via eggs. *Wolbachia* influences some of its insect hosts to produce skewed sex ratios of offspring in favor of females.

VIEWING THE APHID SYMBIONT *BUCHNERA* UNDER A MICROSCOPE

Buchnera is so highly modified that it is not easy to recognize amid crushed aphid tissue under your microscope.

Escherichia Coli and Other Enteric Bacteria

In general, the gamma proteobacteria have done very well for themselves. They include the famous "lab rat" *Escherichia coli*, about which more is known than any other bacterium. When a molecular biologist (or genetic engineer) thinks of bacteria, he or she is most likely thinking of *E. coli*. This species and its myriad relatives are ubiquitous throughout soils, waters, and the digestive systems of most animals. Some species are opportunists and potential pathogens; these include certain strains of *E. coli, Salmonella, Serratia, Legionella*, and *Yersinia* (of black plague fame).

The coliforms, as *E. coli* and its relatives are called, make the news every summer whenever recreational bodies of water become contaminated by sewage. The "coliform count" is the number of these bacteria in a given sample of water. If the count is too high, that beach or pool must be closed until the count goes down again. As the name suggests, *E. coli* and the other coliforms (or enterics) inhabit our colon (large intestine) and are among the many bacteria in our feces; thus, their presence is an indicator of intestinal bacteria (or fecal contamination) in general. As opportunistic gram-negative aerobes, coliforms are fairly easy to identify from field collections. "Coliform" is actually an operational rather than a taxonomic term; however, most coliforms are enterics, a taxonomic group within the gamma proteobacteria that includes *E. coli, Shigella*, and *Salmonella*. Why might it be harmful to ingest a few of our own coliform bacteria? One reason is that the coliforms (and enterics) are *not* on the list of the 25 most prevalent bacteria of the colon and feces; instead, *Bacteroides* and other anaerobes predominate, usually outnumbering *E. coli* and the like 100- to 1,000-fold. In oxygenated waters, however, aerobic coliforms survive longer than the anaerobes. Therefore, a mouthful of coliform-infested water might well disrupt the balance of microbes in the colon, resulting in an unpleasant food poisoning–like infection. (The prevalence of dangerous coliforms in some meat may be a consequence of the misuse of antibiotics in stockyards; see chapter 12.)

It is the ease with which *E. coli* can be isolated and cultured that has made it the best understood bacterium in the world. All its genes have been sequenced, with the workings of many of these genes providing models for understanding other genes. Indeed, culturing *E. coli* is so commonplace that it can be treated almost like any chemical off the shelf (or out of the freezer). It is possible that some molecular biologists have never even seen these organisms in spite of using them every day. Their labs almost never have any microscopes.

Other than the periodic announcements of "coliform events" in the newspaper and your general good health when a reasonably low number of enterics occupies your colon, *E. coli* and its relatives have no particular distinguishing field marks. Even a swimming pond rich in coliforms doesn't look much different from a clean one—except indirectly, in that it may smell of sewage and have algal blooms.

Other well-known enterics (all gammas) include *Salmonella*, famous for contaminating warm food and causing food poisoning. But this species is also an inhabitant of our intestines and is therefore safe in reasonable numbers. An enteric bacterium that was important in human history is *Yersinia*, the cause of black plague and the death of a quarter of the population in Europe during the Middle Ages. *Serratia* is a bright-red bacterium, common in water and soil and a favorite in microbiology classes because of its colorful colonies on a petri dish. It is sometimes visible as red spots on rotting food, but usually *Serratia* and other bacteria are obscured by fast-growing fungi (molds and yeasts).

Delta Proteobacteria: *Myxobacterium*

If you were a science fiction writer with an appreciation of the bacterial world, you still might not come up with something as strange as *Myxobacterium*. For part of their life cycle, these bacteria go completely unnoticed—just one of many gram-negative rods gliding through the soil, digesting whatever organic matter is to be found. In the other part of their life cycle, they are so *un*-bacteria-like that 19th-century botanists used to mistake them for fungi (figs. 8.2a, b). Thousands of them come together and cooperate to make gigantic "fruiting structures" as much as half a millimeter tall—of gigantic proportions in the world of bacteria. Furthermore, their structures are charming in both shape and complexity (fig. 8.3). If you already enjoy the Lilliputian world of lichens, mosses, slime molds, and molds, then you definitely will want to know about myxobacteria. Their life cycle is one of the many strategies that exist for coping with the challenges of life in the soil. (For other strategies, read about the actinomycetes in chapter 11.) In the soil, there is competition for scarce food; by growing slowly and gliding about in films of moisture, myxobacterial rods can find and use bits of decaying plant matter. Under very low nutrient

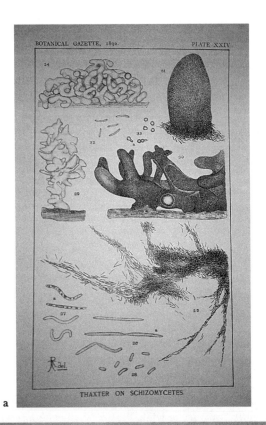

a

Myxococcus aureus Th.
On decaying leaves. n. sp.

Coll. R. Thaxter
Det. "   "   In Lab. cult.
Accession no.   4503   April 1893.

Myxococcus
aureus n.s.
On decaying leaves
in lab. cult.
April 1893

b

FIGURE **8.3.** A diversity of myxobacterial fruiting structures. The tallest structure shown is about 0.75 mm.

conditions, the starved cells band together by the hundreds and thousands to form tall, aerial structures that hold tiny spores high in the air. Water or wind then carries these spores away, making it possible for some to land in more nutrient-rich areas.

Look for these fruiting structures on decaying plant material, the dung of herbivores (especially wild rabbits), and live or dead bark, especially of trees low in tannins and resins (e.g., elders, beeches, and black locusts). Myxobacteria can be yellow, orange, or reddish and can appear either shiny or crusty.

VIEWING *MYXOBACTERIUM* UNDER A MICROSCOPE

Try to arrange some of the structures on a slide and view under the lowest power. A dissecting microscope or magnifying glass will do as well. Be aware that fungi and sometimes slime molds can easily

< FIGURE **8.2a, b.** Roland Thaxter was the first to recognize that myxobacteria were indeed bacteria and not fungi or slime molds. His specimens from the 19th century are still preserved in Harvard's Farlow Herbarium. Schizomycetes, an early term for bacteria, is no longer used.

out-compete myxobacteria, while tiny arthropods may consume them. Myxobacterial structures are smaller than the comparable structures of a fungus or a slime mold. Also, a fungus-infested log is likely to be full of white threads that are visible to the naked eye. In the case of slime molds, their most visible structures are usually dark and crusty, while the mxyobacterial structures tend to be orange and shiny.

CULTURING *MYXOBACTERIUM*

To cultivate these bacteria, you must set bait for them and try to lure and trap them. Here is the basic procedure. Sterilize some wild rabbit dung. Short of wild rabbit dung, you might try domestic rabbit or horse dung. If you do not have a lab sterilizer (autoclave), put some dung in a boiling bag and boil it for about 20 minutes. Half bury some of the dung in a shallow dish of soil. Cover it, and keep the soil moistened. Look daily for tiny structures on the dung. Be patient and vigilant (plate 35).

SUMMARY: FIELD MARKS AND HABITATS OF GAMMA AND DELTA PROTEOBACTERIA

Gammas
- glowing, decaying marine fish or crustaceans
- some glowing terrestrial organisms (e.g., dead caterpillars)
- bioluminescent digestive tracts and other organs of some marine animals (see table 8.1)
- garden frost at temperatures slightly above freezing; some snowfall consisting of especially light, dry flakes (seeded by *Pseudomonas* and *Erwinia*)
- digestive tracts of shipworms (*Teredo*)
- aphids (symbiont: *Buchnera*)

Deltas
- "fruiting" structures on dung, bark, or decaying plant material (*Myxobacterium*)

CHAPTER 9

Gamma and Delta Proteobacteria of Sulfur-Rich Environments

SULFUR-RICH COMMUNITIES

Among the most fascinating of ecological communities are sulfureta (singular: sulfuretum), which are communities rich in hydrogen sulfide. This compound provides a source of energy for the various autotrophic bacteria called sulfur or sulfide oxidizers. Some of these autotrophs produce their own food, as photosynthesizers (or *photo*autotrophs) do, but in this case chemical bonds of sulfide molecules (rather than light) provide the energy source. These autotrophs are called *chemo*autotrophs or (more elaborately) chemolithoautotrophs (for more on chemolithoautotrophs, see chapter 7). In some areas, quantities of sulfide are great enough to support large populations of sulfide oxidizers, which in turn support other organisms, including animals. Just as photosynthesizers form the base of a food web in most sunlit communities, so do the sulfide oxidizers in sulfureta. In completely dark environments such as the deep sea, sulfide oxidizers may be the only synthesizers of food.

The seemingly unusual metabolism of chemoautotrophy that uses the energy of sulfide bonds is rather common among the proteobacteria. Sulfide oxidizers may also be found among the alpha, beta, delta, and epsilon proteobacteria. However, the gamma proteobacteria have the most accessible and visible field marks, and thus are the focus of this chapter. In light environments, sulfide-using photoautotrophs often display lovely pigments; among the most visible of these are the purple sulfurs, which are gamma proteobacteria.

Some of the *producers* of sulfide are sulfate-reducing delta proteobacteria, discussed at the end of this chapter. Their field marks are more olfactory than visual—the production of odoriferous hydrogen sulfide in environments ranging from mudflats to caves to intestines. (I use the terms *sulfate-reducing* or *sulfide-producing* bacteria throughout the book rather than more specific taxonomic names.)

SOURCES OF SULFIDE

There are three major sources of sulfide in sulfureta:

1. Geochemical sources of metal sulfides include springs that carry minerals from deep within the Earth to the surface. Often such springs are thermal.

2. Fermenting bacteria of various types (including gram-positives and proteobacteria) can produce sulfides in the act of digesting foods that are rich in sulfur compounds. These foods include any dead organism that is undergoing decomposition, since sulfur is an important part of the proteins that make up all organisms.

3. Sulfate-reducing delta proteobacteria provide sulfides in some sulfureta. They are in effect "breathing" sulfate and producing sulfide as a waste while breaking down the materials of dead organisms. Sulfate-reducing bacteria and fermenters are often found together in sulfide-rich environments that are based on decomposition.

SULFIDE OXIDIZERS

Microbial communities in which hydrogen sulfide is being produced in abundance are often very poor in oxygen or lack it entirely. Indeed, oxygen reacts quickly with sulfides to produce sulfate, which is the fate of much of the sulfide produced under the relatively oxygen-rich conditions of most of the Earth. Sulfide oxidizers must situate themselves at narrow (sometimes fluctuating) interfaces between sulfide-rich waters or sediments and oxygenated waters or sediments. There seem to be three major "strategies" by which they accomplish this: (1) by associating as symbionts with host organisms in sulfur-rich communities; (2) by being motile; and (3) by being large. These strategies are not necessarily exclusive. For the purposes of making macroscopic identifications in the field, the emphasis here is on large, motile *Beggiatoa* and its relatives, and on relatives of *Thiomicrospira* that are symbiotic in animals.

STRATEGIES OF THIOBIOTIC ANIMALS

The different organisms that inhabit sulfureta are all closely intertwined with each other; so before turning to the bacteria that are the main subject of this chapter, let's take a quick look at the animals and the strategies they use to survive in these habitats.

Environments rich in hydrogen sulfide (e.g., sulfur-smelling tide flats with black sediments) pose particular challenges to their inhabitants. Hydrogen sulfide is poisonous to oxygen-breathing organisms. Just as cyanide binds to the oxygen-carrying molecule hemoglobin, fatally displacing oxygen, so does hydrogen sulfide. The main reason that tidal flats are not lethal to beachcombers is that in spite of the strong smell, there really isn't all that much hydrogen sulfide in the air; most of it reacts quickly with oxygen, becoming harmless sulfate before we breathe it in. We detect the trace that remains because our noses are especially sensitive to sulfidic odors. Animals that live in sulfide-rich environments (sometimes called thiobiotic organisms) have several strategies for avoiding the poisonous effects of this compound.

SPECIAL HEMOGLOBINS

Some thiobiotic animals have special hemoglobins that bind oxygen even in the presence of sulfides. Some have more than one type of hemoglobin—at least one to bind oxygen and one to bind sulfide, which prevents sulfide from binding elsewhere. In some of the segmented marine worms (the polychaete annelids), the gills are conspicuously red with hemoglobin. One polychaete, the blood worm *Glycera*, has a transparent body that reveals hemoglobin throughout. In a few fascinating marine invertebrates, this hemoglobin is indicative of symbiotic sulfide-oxidizing bacteria, as is described later.

MOBILITY

Very tiny invertebrate animals (meiofauna) that dwell among grains of sand are sometimes referred to as "psammophilic" (sand loving). They are difficult to see without special isolation techniques and a microscope. (If you are trying to capture them with a net or filter, these organisms can be retained by a mesh size of 30 to 50 μm.) The meiofauna include worms of all sorts as well as rotifers, gastrotrichs, and gnathostomulids—the last of which are exclusive to sulfidic environments. Not much is known about how the meiofauna thrive in high-sulfide conditions except that they tend to be highly motile. Perhaps they move up and down according to changing conditions. Some are known to feast on sulfide-oxidizing bacteria.

BURROWING

A common strategy for avoiding sulfide is burrowing; the animal's activity keeps the burrow oxygenated with waters from the surface. In this type of environment, it is especially advantageous to be a worm, easily slipping up and down in the sediment. Indeed, most burrow makers are worms of some sort or bivalve (two-shelled) molluscs that have an elongated wormlike shape. Razor clams (*Ensis*) have long shells and bodies and extend a very long "foot" for digging. Others, such as edible soft-shelled clams, or "steamers" (*Mya*), have long siphons by which they exchange water at the sediment surface while remaining buried (plate 36).

The burrows of either worms or molluscs often make good habitats for sulfide-oxidizing bacteria, especially *Beggiatoa* and its relatives, described later. Look for a white haze, film, or scum at the burrow entrance, where oxygen and sulfide interact. Also, fecal pellets ejected from the burrow may become covered with a white scum produced by sulfide oxidizers. The burrows may be lined with these same bacteria, but digging them up disturbs the delicate balance and positions of the microbes.

Chemolithoautotrophic Gammas

One large group of sulfide-oxidizing gamma proteobacteria are chemolithoautotrophs (synthesizers of food using energy in chemical bonds; see chapter 7), described in the following sections. Some of these are symbionts in organisms that inhabit sulfureta or deep-sea thermal vents or that lived inside now-extinct trilobites. Some thrive in whale carcasses and shipwrecks. Others, especially species of *Beggiatoa* and related genera, are found in intertidal areas, sulfur springs, and sulfur caves.

SYMBIOSES IN MARINE ORGANISMS

Certain families of bivalve molluscs, nematode worms, flatworms, and oligochaete annelid worms seem to have taken an additional step in detoxifying sulfide. The mollusc families Lucinidae, Thyasidae, and Solemyidae lack digestive systems but instead have enlarged gills full of sulfide oxidizers—symbiotic bacteria that are closely related to the gamma proteobacterium *Thiomicrospira*. These bacteria, which are chemoautotrophs that use energy from the chemical bonds of sulfides, provide the sole source of nutrition for their host as well as a means of detoxifying sulfide. The lucinids (of the family Lucinidae) are readily identified with any comprehensive guide to shells. Most species are tropical and have round, sturdy, ribbed shells. Two genera of lucinids, *Codakia* and *Loripinus*, sometimes have beautiful pink, yellow, or purple tints on the inner side of the shell (see plate 38).

The worms that cultivate sulfide oxidizers are much more difficult to collect and identify with assurance. In general, if a sulfide-loving worm lacks a mouth and digestive system, it is an excellent candidate for being in a symbiosis with sulfide-oxidizing bacteria. If the worm is white, so much the better, because this is likely due to its bacterial coating (often a source of nutrition for the worm); sulfide oxidizers often deposit white-yellow sulfur compounds. However, it is not easy to confirm that a worm truly lacks a digestive system, since most are skinny, colorless, and elusive, millimeters or less in length. Some nematodes in the subfamily Stilbonematinae, including the genera *Leptonemella*, *Catanema*, and *Laxus*, cultivate an "edible" white coating of bacteria, but it would take a worm specialist to make the identifications.

Some of the annelid worms with bacterial symbionts are in a group, the marine tubificids, that is somewhat more identifiable. They are closely related to the tiny freshwater tubifex worms, available from aquarium shops as fish food, that do not have symbionts but do have red hemoglobin that enables them to tolerate low oxygen conditions. The marine tubificid worms that cultivate sulfide oxidizers and lack digestive systems include *Phallodrilus leukodermatus* and *Inanidrilus leukodermatus* (the identical species name for both alludes to their white coats of bacteria). However, neither these nor other marine tubificids are easily keyed out with ordinary marine field guides.

One type of symbiotic worm that is accessible to amateur naturalists is associated with the tube of a large, easily identified polychaete worm. On many sulfide-rich sandbars, you may be able to find "plumed worms" of the genera *Diopatra*, *Nothria*, and *Onuphis*. At the least, you should be able to see their distinctive tubes. These extend into the sediment as deep as 3 feet and are made of skin-like chitin decorated with bits of broken shells—quite visible at the sediment surface. The enclosed worm is iridescent, has a mane of long red gills, and is about a foot long.

To see both the worm and the full extent of its tube, dig quickly and deeply with a large shovel, turning up as much sediment as possible. Carefully look through the pile, hoping that the tube or the worm itself was not accidently severed (plate 37). (Note that these worms can regenerate to some extent.) To see the mouthless worms with bacterial symbionts (the goal of all that digging) look carefully at the outside surface of a tube. Look for tiny white worms, less than a centimeter long and very thin but often in sufficient abundance to be seen with the naked eye. If you do not see them, try digging up plumed worms from blacker, smellier sediments. The tubes of some plumed worms are host to an entire community of symbiotic worms, including nematodes, flatworms, and oligochaetes, many of them quite tiny and visible only under a microscope.

Colleen Cavanaugh, a discoverer of the deep-sea sulfur communities described next, said this about animals with sulfur bacterial symbionts: they (the animals!) are essentially a means for the sulfur bacteria to "bridge oxic-anoxic interfaces"—a perfectly microbe-centered point of view for this field guide.

DEEP-SEA SYMBIOSES

A remarkable community of molluscs and worms something like the one just described exists in the deepest parts of oceans, such as the Pacific and mid-Atlantic vents. "Remarkable" because until their discovery, nobody had suspected that diverse communities of large invertebrates would be found as much as 3 miles below the ocean surface in dark icy water. Two experts on life without oxygen, Tom Fenchel and Bland Finlay, have noted that much publicity and funding have accompanied the discovery of deep sea–vent communities, while the shallow water counterparts were more or less ignored! However, some of the deep-sea organisms are gigantic (an automatic key to publicity), whereas those of the mud flats are usually tiny.

Thermal vents (underwater hot springs) that spew out sulfide compounds so densely that the water appears smoky are the sole basis of energy for these communities. Sulfide-oxidizing bacteria— the same as those found in tidal flats—are the only autotrophs of deep sea–vent communities, using energy from the sulfide to make sugars. These sugars are in turn consumed, sometimes along with the bacteria themselves, by gigantic molluscs and worms. The molluscs, blood red with hemoglobin, are of the families Mytilidae and Vesicomycidae and are unlikely to show up in ordinary shell collections. The gills of these molluscs are full of symbionts that are similar to the *Thiomicrospira*-related species of more common and accessible molluscs, such as the lucinids described earlier. Huge deep-sea worms, the pogonophorans, or vestiminiferans, are also full of bacterial symbionts and stand upright in chitin tubes, displaying blood-rich tentacles. These worms turn out to be closely related to common annelid worms found in more ordinary thiobiotic environments. Methanogenic symbionts have also been identified in some of these animals, and nitrifying bacteria may be assisting as symbionts in the removal of ammonia wastes.

Once the deep sea–vent communities were discovered, other similarly unusual environments were sought out. "Cold seeps," which are underwater springs of cold, sulfide-rich waters, may support a similar thiobiotic community, although the organisms may be smaller and grow more slowly. One of the more accessible cold seeps, which is also a brine seep, is Gollum's Lake at the East Flower Garden Bank, 100 miles off the coast of Texas. Although 200 feet under water, the seep is essentially a shallow (10 inches deep) salt lake. The sulfide-loving organisms there are fascinating but for the most part quite small, in the range of meiofauna. Among the meiofauna, there seem to be exceptional numbers of gnathostomulids (worm-like consumers of sulfide-oxidizing bacteria). White films and mats of sulfide-oxidizing bacteria cover the sediments. The area is too deep to be reached by scuba divers. Researchers use submersibles or dredges to make observations and collect samples.

SULFIDE-OXIDIZING COMMUNITIES OF WHALE FALLS AND SHIPWRECKS

The communities around hydrothermal vents typically use a geochemical source of sulfides for energy. However, alternative sources are sometimes used, especially when a large quantity of organic material suddenly becomes available to bacteria. When a whale dies, for example, its corpse, called a "whale fall," becomes an instant banquet for fermenting bacteria of many types as well as sulfate-reducing bacteria (see the section at the end of this chapter). Both types of bacteria release an abundance of sulfide compounds that often turn the surrounding sediments black with iron sulfide. Although this process is usually a deep-water phenomenon, sometimes a shallow-water whale fall can be viewed. Even if it is quite old—nothing but bones—it may be surrounded by a lot of black, sulfide-rich sediment. Into such a sulfide-rich community come the sulfide-oxidizing bacteria, and sometimes the animals associated with them. Thus, a whale fall can provide an opportunity to study the same sorts of thiobiotic animals that are found in much deeper hydrothermal vents.

Shipwrecks containing cargos of food are another source of abundant nutrients on which a sulfide-generating, sulfide-oxidizing community can develop. For example, a ship loaded with beans and grain sank off the coast of Spain and became the basis of a community that supported animals similar to those hydrothermal vents. Such ships can also provide a substrate for iron-oxidizing bacteria (see chapter 7).

TRILOBITES AND SULFIDE-OXIDIZING SYMBIONTS

Trilobites are extinct and can be studied only through their fossils. At one time, they were extremely successful, abundant, and diverse dwellers of the ocean floor. According to a hypothesis of paleontologist Richard Fortey, one tribolite family, Olenidae, may have contained sulfide-oxidizing symbionts. The trilobites most likely lived in sulfide-rich environments: they have reduced mouth parts, and their oxygen-exchange organs are enlarged in a way that might accommodate bacterial symbionts.

Slate is a sedimentary rock sometimes formed from deep, sulfury sediments of ancient thiobiotic communities such as the ones in which trilobites once thrived. Slate that was formed with abundant organic matter is black, whereas that formed in the presence of iron in oxygen-poor waters is greenish-purple. If oxygen is available, the iron in slate can yield a rusty red color. You can see traces of these ancient sediments (that very likely once supported a thiobiota) by looking for slate as a building material, gravestone, countertop, or chalkboard.

BEGGIATOA AND RELATED SULFIDE OXIDIZERS

Sometimes an estuary, hypersaline environment, or sandy intertidal area shows a visible enrichment of *Beggiatoa*. Species in the genus *Beggiatoa* and in related genera (called here "*Beggiatoa*-types") oxidize hydrogen sulfide to sulfur, accumulating the latter as a waste product. In doing so, they gain energy and hydrogen with which they can make sugars (food). *Beggiatoa* is similar to photosynthesizers, but it is actually a chemolithoautotroph because it gets energy from chemicals—in this case, hydrogen sulfide. (For more on this metabolism, see chapter 7.) Some *Beggiatoa* species, however, seem to be heterotrophs and use sulfide oxidation as a detoxification mechanism.

To observe *Beggiatoa*, look (and smell) for black hydrogen sulfide–rich sediments produced by sulfate-reducing bacteria. *Beggiatoa* often produces a white scum or a cobweb-like array in close proximity to the black sediment; if black sediment is not near the surface, you may have to gently look under a layer of photosynthesizing bacteria, which may form either a green mat or merely a green discoloration of the substrate. *Beggiatoa* will not be deep in the sediment because it requires oxygen. Its optimal location is at the interface of anaerobic, hydrogen sulfide–rich sediment (or water) and oxygen-rich sediment (or water) (plate 38).

Animal burrows and decaying seaweed in some marine environments are also reliable sources of *Beggiatoa*-type bacteria. Burrows in sulfide-rich sediments often form interesting interfaces at the burrow entrance between the oxygenated waters above and the sulfide brought up from below. The worms or molluscs of such burrows sometimes have sulfide-oxidizing symbionts, but they may also create conditions for looser associations with *Beggiatoa* types that sometimes proliferate at the burrow entrances.

Decaying seaweed can provide a wonderful temporary sulfur community. Fermenters that break down the seaweed tend to exclude oxygen from their immediate area and to release sulfur compounds from the seaweed. These compounds in turn are used by sulfide-oxidizing *Beggiatoa* types, which form a film or coating or even fine, feathery, white dusting that can sometimes be observed on seaweed (plate 39). In some situations that are not well understood, a fine coating of sulfide-oxidizing bacteria may be part of the normal life cycle of a seaweed (this may be the case with the brown, filament-like seaweed *Chorda*).

If you are in a part of the country that has cool to warm sulfur springs (providing a geological source of hydrogen sulfide), you may have an opportunity to see massive enrichments of *Beggiatoa*. (For a discussion of sulfur springs hotter than about 75°C [167°F] see chapters 1, 3, and 13.) Those long, filamentous bacteria, white with accumulated sulfur, love to be in the vicinity of hydrogen sulfide, which you can detect as the smell of rotten eggs. Sometimes they form white mat-like layers or tufts, or even puff balls of white filaments that are visible to the eye (plate 40). By the way, it may not always be apparent whether a source of sulfide in a spring is from sulfate-reducing bacteria or from a geochemical source; some bacteria can be active deep in the ground.

Beggiatoa types may also be found on the walls of sulfide-rich caves; in sulfur springs, including deep-sea springs; and indeed in almost any environment where sulfur is dominant. Other possibilities include sewage pipes, eutrophic bodies of water, rice paddies, and stands of *Spartina* in marshes.

Size and Motility in *Beggiatoa* and Its Relatives

Beggiatoa and its relatives provide good examples of two strategies found in many sulfide-oxidizing bacteria: large size and high motility. These two traits make *Beggiatoa* types among the most easily observed of all the bacteria. Its long, thick, multicellular filaments enable *Beggiatoa* to span the zones of sediment in which oxygen and sulfide are both optimal. An elongated form may allow these bacteria to sense subtle changes in oxygen and sulfide conditions along this interface and to respond accordingly. Response is in the form of fast, gliding movement—up to 150 μm per minute. In fact, *Beggiatoa* cells are constantly on the move, seeking exactly the right interface. They are well known to migrate up and down in sediments or water according to tide changes and on day–night cycles. At night, when photosynthesis and oxygen production shut down, *Beggiatoa* (and also pink *Chromatium*, described later in this chapter) glide (and swarm) up through the water column even to the point of making the water whitish or pink. The sulfide smell is stronger on a tide flat at night for this same reason—the shutdown of photosynthesis. You can do simple daytime experiments by blocking the light from *Beggiatoa* and cyanobacteria, thereby inhibiting photosynthesis and encouraging *Beggiatoa* to glide up.

Sulfide oxidizers are among the largest bacteria, the huge sphere-like *Thiomargarita* (a relative of *Beggiatoa*) perhaps taking the prize for the biggest prokaryote. But consider this: most of *Thiomargarita* is taken up by a huge bubble-like vacuole of nitrates, which it uses to oxidize sulfide; it has relatively little cytoplasm (cell contents). In a sense, *Thiomargarita* can "hold its breath" by storing nitrate rather than adjusting its position to optimize oxidation. *Thiomargarita*, found off the coast of Namibia, is not particularly accessible; however, *Thioploca*—which also has huge vacuoles and relatively little cytoplasm—can be found in some of the same environments as *Beggiatoa*. It consists of long, thick, multicellular filaments that are "braided" together in a common sheath.

SULFUR CAVES AND SCHOLARS' ROCKS

Most caves of sedimentary rocks are formed in karsts, which consist of carbonate rocks such as limestone or, less commonly, evaporite rocks such as gypsum. Karsts are the sediments of ancient seas, although modern-day karstic settings often bear no resemblance or proximity to seafloors as a result of continental drift and other geological changes. Caves are usually formed when karstic rocks are dissolved by carbonic acid, an acid that is formed easily by carbon dioxide and water.

Some caves are formed in karsts by sulfuric acid, the product of sulfide-oxidizing bacteria that use hydrogen sulfide as a source of energy. In this case, the source of hydrogen sulfide is sulfate-reducing bacteria. True "sulfur caves," which have active sulfur cycles mediated by bacteria, are rare enough that even most devoted spelunkers have probably not visited one. These caves tend to be quite uncomfortable and may even be dangerous due to potentially lethal levels of hydrogen sulfide and the presence of dripping, burning sulfuric acid. Cave explorers use gas masks and protective clothing to explore such environments. Even then, the gases and acids can eat away at material and corrode equipment. Although ordinary karstic caves can sometimes have areas rich in sulfurous minerals such as gypsum, there are only three major, active sulfur caves in the world: Lechuguilla in New Mexico, Cueva de Villa Luz in Tabasco, Mexico, and Movile in Romania. All three are more or less off limits except to researchers.

Lechuguilla Cave is in Carlsbad Caverns National Park, but it is not open to the public because of concern that its unusual and delicate (and exceptionally beautiful) formations might be destroyed. (For an interesting account of research in this cave, see the March 1991 issue of *National Geographic*.) Sulfide-oxidizing organisms are active in the cave, producing sulfuric acid and contributing to the slimy cave surfaces and the caustic air. There is good evidence that most or all of the caves of Carlsbad Caverns, including those open to tourists, are products of the dissolution of sulfuric acid. Therefore, although a sulfur cycle is no longer active in the main cave, the resulting spaces through which you can walk are products of past bacterial activity.

A more accessible cave system that has sulfide springs associated with it is Parker Cave in Kentucky. Phantom Waterfall and Sulfur River of that same cave system have white mats of filamentous, sulfide-oxidizing bacteria.

"Scholars' rocks" are fantastically shaped and perforated carbonate (and other) rocks that have been collected and revered for thousands of years by the Chinese. Ranging in size from paperweights to huge boulders, they resemble the cave-riddled cliffs and gorges often depicted in Chinese paintings. Sulfide-oxidizing bacteria, such as those just described from caves, are essential sculptors of scholars' rocks because they provide the acidic compounds necessary to form the shapes and holes.

VIEWING *BEGGIATOA* AND OTHER SULFIDE OXIDIZERS UNDER A MICROSCOPE

Scrape some white sediment from the sand or seaweed, or the white coating from a thiobiotic animal. Look for long, clear filaments of bacteria that have bright-white sulfur granules inside.

CULTURING *BEGGIATOA* AND OTHER SULFIDE OXIDIZERS

It is easy to culture *Beggiatoa* using a Winogradsky column (see appendix A and the section in this chapter on culturing *Chromatium*), but first look at the next section on purple sulfur bacteria, because you will most likely be culturing these as well. See also the chapters on cyanobacteria (chapter 13) and on green sulfur bacteria (chapter 5) and the section on sulfate-reducing delta proteobacteria (later in this chapter); these too will flourish in your enrichments.

A modification of the Winogradsky method for enriching *Beggiatoa* and other sulfide oxidizers was devised by sulfuretum expert Bo Jørgensen. Fill a glass loaf pan with about 1 inch of sand, 1 inch of black sulfur mud, and a few inches of seawater diluted by about 50% with fresh water. Add a few dead leaves or hay as well as a sprinkle of Epsom salts as a source of sulfate, unless you have access to calcium sulfate, as suggested by Jørgensen. Wrap half of the loaf with foil, and put the unwrapped end near a light source such as a window. After a few days or weeks, you should be able to observe cobwebby, filmy, white sulfide oxidizers and other sulfuretum bacteria on the dead vegetation and sediment surface. A quicker but smellier preparation entails putting a piece of recently dead fish or mollusc into a jar of seawater and waiting a day. You may be rewarded with white clouds and films of sulfur bacteria before putrifying fermenters settle in and cause you to throw the mixture away.

PHOTOAUTOTROPHIC GAMMAS: PURPLE SULFURS

Many of the bacteria described in this chapter either live in darkness, are active primarily in habitats into which little light penetrates, or move in and out of the dark. When sulfur communities do occur in photic zones, however, wonderfully colored photosynthesizing bacteria may be highly visible: these include the purple sulfur bacteria.

CHROMATIUM

Chromatium and its relatives are photosynthetic gamma proteobacteria that thrive wherever there is hydrogen sulfide in a photic zone. They are readily identified by the lovely pink-purple colors that give the purple sulfurs their nickname. In fact, any pink or purple coloring found in the vicinity of hydrogen sulfide is almost guaranteed to include *Chromatium* and related genera (plates 41, 42). But don't confuse these with the red-pink halobacteria (chapter 4), which grow on salt crystals or salt crusts in extremely saline environments, or some of the purple sulfurs of alkaline environments, which are described in the next section. *Chromatium* is a photosynthesizer, but it does it a little differently from cyanobacteria (and green plants and algae). While cyanobacteria, plants, and algae use water as their source of hydrogen, purple sulfur bacteria use hydrogen sulfide. *Chromatium* is most commonly found in contact with black, anaerobic, sulfide-rich sediments that are still within the photic zone. Often they are just underneath a layer of blue-green cyanobacteria. The purple-pink pigments of *Chromatium* are a type of carotenoid that absorbs light and passes the energy on to the chlorophyll. Essentially *Chromatium* scavenges light that is not absorbed by the cyanobacteria, taking advantage of wavelengths that penetrate through the bacterial mat above it.

The position of *Chromatium* beneath cyanobacteria is analogous to the layering of some seaweeds. Green seaweeds are often positioned to capture light in shallow water. Brown seaweeds may be found attached to rocks below the greens, and reds are found even deeper. Red seaweeds, such as *Chondrus crispus*, are able to capture light that penetrates deep into the water by using reddish secondary pigments.

Chromatium bacteria are highly motile and are known to swarm up from the sediment in a pink cloud when oxygen-generating photosynthesis has shut down for the day. A stronger sulfur smell at night may therefore indicate the activities of *Chromatium*.

Look for purple sulfurs in beach sulfureta, estuaries, and mudflats where sediments are stable enough to form a surface layer. Gently remove the cyanobacterial layer or cut down through the sediment to reveal a cross-section. The pink-purple layer may be on the surface if hydrogen sulfide is abundant there. Sulfur springs and some freshwater systems with sulfur-rich sediments (e.g., karstic lakes and ponds or waste stabilization ponds) also show enrichments of purple sulfurs. These bacteria can be difficult to see in a deep, stagnant body of water because they may form a layer several centimeters below the surface. If sulfide is abundant enough, however, sometimes a body of fresh water is pinkish all the way to the top.

VIEWING *CHROMATIUM* UNDER A MICROSCOPE

Place some of the pink material that you believe to contain bacteria on a slide, removing by hand any sand grains that will prevent you from making a flat preparation. Look for rods or cocci, often plump and full of bright-white sulfur deposits and sometimes gas vacuoles. Some rod-shaped purple sulfurs are highly motile and can look like little pink mice scurrying around (fig. 5.2).

CULTURING *CHROMATIUM* AND OTHER BACTERIA OF SULFURETA

When you culture *Chromatium* in a Winogradsky column (see appendix A for general instructions), you will also be culturing cyanobacteria, *Beggiatoa*, sulfate reducers, and a host of other organisms (plate 43). Use a large jar, as tall and column-like as possible, although ordinary mayonnaise and peanut butter jars will work. Place a little torn-up paper, rice, seaweed, beach grass, or some other source of cellulose in the bottom, and add a few spoonfuls of calcium sulfate ($CaSO_4$) and a pinch of potassium phosphate (K_2HPO_4). If those chemicals are not handy, use 2 to 3 spoonfuls of Epsom salts (magnesium sulfate) from the drugstore. By adding minerals and cellulose, you are trying to mimic the effects of having sulfate-rich rocks or sediments and lots of decaying organic matter in an anaerobic environment. You can be creative with which "food" you select.

Try something different from cellulose to see what happens. Egg yolks, for example, are rich in sulfur and can produce good results.

Next, scoop in black, sulfury sediments. Don't worry that you are turning the microbial habitats all topsy turvy; they will sort themselves out. Add water from the environment so that a centimeter or two covers the top. Decide at what vertical height you would like the photic zone to be located, then cover everything else with black paper. (Black paper taped two-thirds of the way up the jar is a good place to start.) Place your completed column in a north window, unless you would like to experiment with using direct sunlight in a south window. You can either tighten the lid or leave it loose, but be aware that a loose lid will sometimes release malodorous hydrogen sulfide if the community is not balanced.

One of the first things that will happen is that sulfate reducers (sulfur-breathing heterotrophs, described later in this chapter) will begin to consume the fermenting cellulose products and produce hydrogen sulfide as a waste product. In the photic zone, a pinkish layer of purple sulfurs will form and above that a blue-green layer of cyanobacteria (chapter 13). Sometimes a layer of green sulfur bacteria (chapter 5) will be situated below the pink layer. You may also find that below the pink and perhaps below the edge of the black paper, white *Beggiatoa* may develop. If you are not getting *Beggiatoa*, a little oxygen may help. Consider making another column in which you place an air stone attached to an aquarium air supply, set very low. *Beggiatoa* will seek out the optimal position between the air stone and the black sulfide layer. Keep in mind that cyanobacteria are also sources of oxygen, since it is their waste product; *Beggiatoa* may situate itself strategically nearby.

Occasionally add a little tap water to the column and turn it so that a different side faces the window. Enjoy it for years! It makes a great conversation piece for the kitchen or classroom windowsill. If all is going well, a complete sulfur cycle should be taking place within, so there should be no smell of hydrogen sulfide unless you dig into it.

PURPLE SULFURS OF ALKALINE LAKES

Chromatium and its purple sulfur relatives are most likely responsible for any pink-purple coloration of sediments or waters if sulfides are present. However, a different group of purple sulfurs, spiral-shaped bacteria in the family Ectothiorhodospiraceae, can be found making pink-purple blooms in soda (sodium carbonate), or alkaline, lakes. To find these bacteria and other alkaline-loving (alkalophilic) organisms, find an alkaline lake or spring, often designated by the word *soda*, such as Big Soda Lake in Nevada. Like any purple bacteria, Ectothiorhodospiraceae seek out a special layer within the photic zone but deep enough to be in contact with the hydrogen sulfide that is generated farther down. Thus the bloom may be many meters (as much as 20) deep. Shallow waters such as alkaline springs may be more promising (plate 44).

Other bacteria to be observed at soda lakes include alkalophile sulfate reducers (see next section), the generators of all that hydrogen sulfide. Also, methanogens (see chapter 2) may be detected by bubbles of methane gas rising to the surface. Where there are methane producers, there are usually methane users (such as methane oxidizers), but these are not directly observable. Halophiles that tolerate extreme alkalinity may also be present (see chapter 4).

SULFATE-REDUCING DELTA PROTEOBACTERIA

The sulfate-reducing delta proteobacteria are famous for producing the smell of rotten eggs (hydrogen sulfide) that often pervades salt marshes, mudflats, and other areas with dark anaerobic sediments (plate 45). These same deltas are responsible for some of the trace sulfur-rich gases that can cause intestinal gas to be malodorous. Thus, it is by no means coincidental that low tide and flatulence can be hard to distinguish.

Almost anywhere, almost anytime, the smell of hydrogen sulfide (a sharp, sulfury, rotten-egg smell) is the definitive field mark of the sulfate-reducing bacteria. An exception is some mineral-rich spring waters from deep in the ground. Although sulfate-reducing bacteria can be active quite deep in sediments, if these waters are from an anaerobic source, then the sulfide may have a nonbiological origin.

Sulfate-reducing bacteria prefer environments with materials that are tough to digest and rich in cellulose, such as decaying plants or seaweed. Other bacteria ferment the cellulose slowly, releasing compounds that the sulfate-reducing bacteria can digest. Sulfate-reducing bacteria also obtain sulfate from seawater, from sulfate minerals in the sediments, or from wastewaters that are rich in organic materials. Just as we and many respiring heterotrophs breathe oxygen, sulfate-reducing bacteria breathe sulfate. The waste product of oxygen respiration is water, while that of sulfate-reducing bacteria is hydrogen sulfide.

MARINE WATERS AND SEDIMENTS

Look for sulfate-reducing bacteria in black-gray, smelly muds of estuaries, salt flats, and other marine sediments. The blackness is due to a reaction of hydrogen sulfide with iron-forming black ferrous sulfide (FeS) or gray pyrite (FeS_2). These black sulfide minerals and the smell of sulfide are field marks of the sulfate reducers.

FRESH WATERS AND SEDIMENTS

If a body of fresh water smells of sulfur (and is unpolluted), it may be karstic, which means that at some point in its geological past it was part of a body of salt water, now evaporated. Karstic lakes are usually rich in carbonates. More rare are the sulfur-rich gypsum/anhydrite systems in which sulfureta can form. Sulfur-rich karstic lakes of the United States include those of Bottomless Lake State Park in New Mexico. Such an environment is often quite porous and is full of carbonate caves and deep wells. Sulfate-rich minerals may dissolve in the water, providing a resource for sulfate-reducing bacteria. A sulfide smell is their field mark. Because lakes of this type have a photic zone, expect to see sulfide-using photo-synthesizers (the purple sulfurs described earlier) coloring the water pink during some times of the year.

Some ancient carbonate sediments have become converted to dolomite (calcium magnesium carbonate) by processes not well understood. One hypothesis is that the sulfate-reducing bacteria caused dolomite to be precipitated (deposited) as an indirect chemical effect of using sulfate. If this is correct, then dolomite deposits may be interpreted as macroscopic field marks of ancient sulfate reducers. Some karstic systems with gypsum, such as the sulfur caves described earlier, form sulfureta.

NUTRIENT-RICH WATERS

Any body of fresh water that is rich in decaying organics (e.g., holding ponds for water treatment or eutrophic ponds) might smell like sulfide from the activities of sulfate-reducing bacteria as well as fermenting bacteria. The canals of Venice, for example, are richly polluted with organic materials and sometimes smell of sulfide. The blackness of the iron sulfides stains the gondolas, which are painted black to make the staining less conspicuous.

Overfertilized golf courses sometimes form a black, sulfide-rich layer below the turf as a result of the activity of sulfate-reducers metabolizing the excess sulfates from the fertilizer and zealous watering. Rice paddies may also be rich in sulfates, providing an environment for the deposition of black iron sulfides.

In general, any rich source of nutrients has the potential to be a habitat for sulfate-reducing bacteria. Some especially large and long-lived sources can form the base of complex sulfur communities. For example, see the description of sunken ships and whale falls as sulfur-based communities earlier in this chapter.

VIEWING SULFATE REDUCERS UNDER A MICROSCOPE

Try to look at some black sediment, first removing any tiny grains of sand that might make the preparation bumpy. However, be prepared to find *many* bacterial types. Some common sulfate reducers are vibrios (*S*-shaped bacteria), but it will be difficult to know which vibrios in the preparation are in fact sulfate-reducing bacteria.

CULTURING SULFATE REDUCERS

If you make a Winogradsky column as described in appendix A and earlier in this chapter, you will find it difficult to avoid culturing sulfate reducers. Furthermore, you will discover that several bacterial types that oxidize hydrogen sulfide are easy to culture along with the sulfate-reducing bacteria.

Almost any source of cellulose covered by stagnant water eventually becomes rich in sulfate-reducing bacteria. For example, a bucket of cuttings from forsythia bushes, or any twigs and branches left to blossom in early spring, can provide food for cellulose-digesting fermenters, which in turn provide conditions for the growth of sulfate-reducing bacteria. If you leave the cuttings for a period of weeks, the water will begin to smell of hydrogen sulfide.

SULFATE-REDUCING INTESTINAL BACTERIA

Most intestinal gas is composed of nitrogen, oxygen, and carbon dioxide, the very gases that we swallow in the act of speaking, eating, and drinking. These gases are odorless and colorless, as are the gaseous wastes of many of the myriad bacteria (mostly *Bacteroides* and gram-positives) that inhabit our intestines, fermenting whatever food we cannot process with our own enzymes. These gaseous products of fermentation include more carbon dioxide, hydrogen, and sometimes methane. (For more details on methane production, see chapter 2.) Between 400 and 2,400 mL of these relatively benign gases are emitted by every human every day.

So what smells so bad? It is the trace amounts of sulfur-rich gases that are produced by sulfate-reducing bacteria and some fermenters, especially in response to foods that are rich in sulfur compounds. These include members of the onion and broccoli families as well as the notorious beans and certain beverages, such as beer. Cooking sometimes releases these compounds, which is why overcooked broccoli or cabbage can smell sulfury. Many food preservatives contain sulfates too. Proteins (in foods such as meats and grains) are full of sulfur compounds, which are enhanced on roasting. Also, some of the more interesting cheeses contain sulfur compounds. In fact, when it comes to trace amounts of sulfur compounds, there is a very fine line between a good (even delicious) smell and a bad smell. Some of our most flavorful and aromatic foods are substrates for sulfate-reducing bacteria. The gases they produce include hydrogen sulfide, methanethiol, and dimethylsulfide. Other smelly sulfur compounds include skatole and indoles. A diet rich in delicious sulfury food can result rather quickly and directly in activities of sulfate-reducing bacteria and thus malodorous flatulence.

And then there is flatulence in dogs. With diets far removed from those of ancestral wolves, dogs often consume sulfates in their processed diets and produce malodorous sulfides in their gas. Some breeds are more prone to this than others, perhaps due to genetic differences inadvertently selected by breeders along with more positive traits. Researchers at the Waltham Center for Pet Nutrition in England have determined that a supplement of activated charcoal, zinc acetate, and yucca extract can ameliorate the odor.

Interestingly, the methanogens (chapter 2) and the sulfate-reducing bacteria are in competition for similar substances; if sufficient sulfur compounds are present, the sulfate-reducing bacteria prevail. Studies comparing the intestinal flora of people in industrialized countries with that of people in developing countries have shown that the methanogens predominate over sulfate-reducers in people of developing countries, perhaps because the diet is less processed and less likely to contain sulfate-rich preservatives and flavors.

Hydrogen sulfide is quite toxic to oxygen-respiring organisms. Like cyanide, it binds to the places where oxygen should bind in a respiring cell. If you breathe in a small amount of cyanide, it can be deadly. Why, then, are we not dead from smelling salt marshes or flatulence? Perhaps because we are so sensitive to the smell of sulfur; we smell hydrogen sulfide and avoid it far before it reaches toxic levels. Nevertheless, constant exposure to hydrogen sulfide can be damaging. One of the more concentrated sources of this gas can be the intestines of people in which sulfate-reducing bacteria are very active. In constant, large quantities, the gas may cause or exacerbate diseases of the colon.

SULFATE REDUCERS AND METAL OXIDIZERS
Sulfate-reducing bacteria sometimes interact with iron-oxidizing bacteria in iron water pipes, resulting in corrosion of the pipes and a sulfide smell in the water (see also chapter 7). The iron oxidizers find a natural substrate in the iron of the pipe, producing oxidized iron or rust. As the bacteria live, thrive, and then die, they leave dead cells and wastes that adhere to the piping. This debris is a source of organic material and sulfate for sulfate-reducing bacteria. As sulfate reduction occurs, the waste product sulfide reacts with the iron to form iron sulfides and localized corrosion and pitting of the pipe.

If other metals, such as copper and zinc, are available, they may form sulfides too. If there is pollution from overly rich nutrients and metals, sulfate-reducing bacteria may actually stabilize the metals, in effect "cleaning" them up.

SUMMARY: FIELD MARKS AND HABITATS OF GAMMA AND DELTA PROTEOBACTERIA OF SULFUR-RICH ENVIRONMENTS

(*Note*: *Beggiatoa* and *Chromatium* are used here as representative genera for white sulfur-oxidizing bacteria and purple sulfur bacteria, respectively.)

Gammas

Marine environments
- white layer or scum in close proximity to black, sulfur-smelling sediment or under green mat of photosynthesizing bacteria in estuaries and tidal flats (*Beggiatoa*)
- white film, scum, or tufts on decaying seaweed or on fecal pellets around worm or mollusc burrows in sulfur-rich sediments (*Beggiatoa*)
- some molluscs, nematodes, polychaete and other worms in tidal flats (bacterial symbionts)
- Pink-purple scum exposed on the surface or just beneath a layer of blue-green-black cyanobacteria in sulfide-rich estuaries and intertidal flats (*Chromatium*)
- deep-sea vents
- some whale carcasses or nutrient-rich cargoes of sunken ships (sulfur bacteria)
- some trilobites

Sulfur springs (cool to warm)
- white scum, tufts, or puffballs of white filaments (*Beggiatoa*)
- pink-purple scum (*Chromatium*)

Ponds, lakes, springs
- pinkish coloration, sometimes occurring seasonally, in karstic lakes (*Chromatium*)
- purple bloom in alkaline (soda) lake, sometimes very deep, or in alkaline spring (purple sulfurs in the family Ectothiorhodospiraceae)

Sulfur caves and scholars' rocks
- white, filamentous mats in caves (*Beggiatoa*)
- holes and perforations in rocks

Other habitats
- pink coloration in a waste pond (*Chromatium*)

Deltas (Sulfate-Reducers)
- gray to black sulfur-smelling mud of estuaries, salt flats, and other marine environments; some slates
- deep waters and sediments of freshwater karstic ponds and lakes, water-treatment holding ponds, or other freshwater environments where there is considerable decay of vegetation
- intestinal gas that smells sulfurous (note that ordinary fermenting bacteria in the digestive system, such as gram-positives, may also produce sulfurous gases)

Introduction to the Gram-Positive Bacteria

The gram-positive bacteria are named for their tendency to be irreversibly dyed blue by crystal violet stain. This stain, or dye, discovered in 1884 to be effective on bacteria by the Danish physician Christian Gram, is now called Gram's stain. Any bacteria permanently stained by it are considered to be gram-positive.

What Christian Gram could not have known was that this particular stain, of all of the hundreds of stains used for visualizing bacteria, would become one of the essential methods for identifying almost any bacterium. It turns out that most bacteria fall into one of two large groups: those that stain with Gram's stain—"the positives"—which have a thick but relatively simple cell wall; and those that do not—"the negatives"—which have a thinner, more complex wall and an outer membrane (plates 46a, b).

A few bacterial groups do not fall neatly into either of these categories. The "gram-variables" show inconsistent responses to Gram's stain, but when examined closely with the electron microscope, they seem to be variants of gram-positives, based on the appearance of the cell walls. Other bacteria have no walls at all and are usually found inhabiting other cells. These, too, turn out to be gram-positives that lost their walls when they evolved as symbionts of other cells. The archaea (archaebacteria), which are different from other bacteria in myriad ways, are also different in their cell wall structures. Gram staining is not used to identify the archaea.

More than a century after Gram's discovery, DNA sequences provided evidence that gram-positive bacteria (along with the gram-variables and secondarily wall-less types) form a cohesive and very successful group comprising its own major branch of the bacterial family tree. The other nine main bacterial branches are gram-negative. It would be convenient to attribute the success and near ubiquity of the gram-positives to their cell walls. However, the gram-negative bacteria as a whole are also extremely successful and nearly ubiquitous. It is not apparent that one or another type of cell wall confers special advantages in any particular environment. Some gram-positives can withstand dry spells, possibly as a result of their thicker walls. However, many other gram-positives never experience drying, and many gram-negatives have evolved their own adaptations to resist drying. Gram-positives and gram-negatives are often found together in the same environments, both apparently well adapted.

Gram-positive bacteria presently occupy nearly every conceivable ecological niche, ranging from fresh water, salt water, sediments, and soils to the surfaces and insides of other organisms. Many gram-positives are heterotrophs, digesting a wide variety of food molecules in their environments; one group of gram-positives can photosynthesize, making its own food.

A few gram-positives are opportunistic pathogens: if the right circumstances arise, they can invade and harm other living organisms. See the section on pathogens in the general introduction for a brief review of what makes a small percentage of bacteria invasive and harmful. In fact, the original tests of the Gram-staining procedure were made at a time when researchers were becoming more and more aware of bacteria as agents of disease. Stains were being developed to make tiny, colorless bacterial cells more visible among the diseased animal tissues from which they were isolated. Gram was originally disappointed with his stain because only some of the bacteria became colored. He had hoped to visualize all of the bacteria in his samples.

The gram-positives are an extremely diverse group. Their cells come in almost every conceivable shape and size (within the confines of bacterial possibilities). Many are simple rods or cocci and, as such, are extremely difficult to distinguish from rods and cocci from other bacterial groups. Some are arranged in long filaments; some form helical shapes; some wall-less forms are quite amorphous. Many gram-positives form tiny spores that are resistant to drying and starvation. This strategy has enabled them to be very successful in environments that are subject to periodic losses of nutrients or water.

Major Groups of Gram-Positives

Taxonomists using DNA sequence analyses have found that gram-positives may be divided neatly into two groups: The "high-GC" gram-positives (also called actinomycetes) and the "low-GC" gram-positives. "GC" refers to guanine and cytosine, two of the four bases that make up most DNA sequences. The other two bases are adenine and thymine (A and T). High-GC gram-positives have somewhat more G and C in proportion to A and T than the low-GC gram-positives do. The significance of this is not well understood, however, and determining GC ratios is not a routine procedure in most microbiology labs. Unfortunately, there is no bit of bacterial etymology or lore that makes these names especially memorable.

Because many of the field marks described in these chapters include complex communities of both high-GC and low-GC types, these terms appear primarily in the chapter summaries, for reference, rather than in the text descriptions.

Field Marks of Gram-Positives

The gram-positives we are most likely to identify by their field marks are those that we encounter as enhancers of flavors and aromas in our foods and drinks. Chapter 10 is devoted to cheeses, pickles, wines, beers, and other delicious bacterial by-products.

A few of the gram-positives of soil and other substrates also have distinctive field marks (chapter 11). Some nonpathogenic gram-positives of our skin and body cavities can also be distinguished by field marks. These and other symbiotic gram-positives are the focus of chapter 12.

CHAPTER 10

Gram-Positive Bacteria of Foods and Drinks

There is a fine distinction between food enhanced by bacteria and food spoiled by bacteria. Often the former is an acquired taste, as bacteria can confer dramatic changes in odor, flavor, and texture. A certain amount of acidic, distinctively aromatic, volatile compounds, and even slime may be desirable; but too much of any of these can be nauseating and even near toxic. Perhaps hunger and even near starvation in our human ancestors were essential to the evolution of great cuisines. In turn, great cuisines themselves may have been influential in the evolution of human cultures in that specific preferences for certain flavors, aromas, and textures can be clear definers and even maintainers of "our culture" versus "theirs." No matter how open-minded an adult might be about another's customs, the line is often drawn at consuming with true enjoyment (and with lack of digestive distress) another's fermented food. Thus, bacteria, as producers of some of our most culturally specific foods, deserve credit for their role in defining the boundaries of many of our diverse human cultures.

The readers (and writer) of this book are likely to be well fed and even a bit adventurous with sharp cheeses, raw fish, and wild mushrooms. However, we are unlikely to taste spoiling food that is frothing, turning colors, and producing strange odors in the hopes that it might be still edible. That is where true hunger (not just appetite) plays a crucial role in the evolution of edibles. Our adventurous (and hungry) ancestors probably got their share of upset stomachs and food poisoning. Even into the 19th century, housewives were being given advice as to how to reclaim foods that had become musty or rancid. For example, Lydia Marie Child and Catherine Beecher include such methods among their recipes. However, the author of this book is not recommending casual experimentation with rotting food!—pour that bottle of spoiled milk down the drain!

Lactic acid–producing gram-positives, called lactobacteria, are literally everywhere, seemingly poised to drop into food. Often their effect is undesirable—inappropriately tart tastes, odd smells and flavors, scum, and slime. The afflicted food item may be considered to be "rotting" in its most negative sense and ought to be tossed out. As with many delicious cheeses and beers, however, the presence of lactobacteria in food can enhance it and can be a preservative of sorts; once lactobacteria and their acidic wastes are established in a food item, other bacteria are less able to colonize. The great cuisines of the world all include some form of bacterially produced, pickled, acidified, and piquantly flavored foods, a few of which are listed in table 10.1.

What lactobacteria are doing is fermentation, a type of heterotrophic metabolism found in nearly every branch of the bacterial family tree as well as in many fungi. (Although fermentation is not the sole supporting metabolism for animals like ourselves, we do it too, and on brief occasions when parts of our bodies are deprived of oxygen, fermentation predominates. Lactic acid, a waste product of fermentation, sometimes builds up in exercising muscles, causing cramps until conditions are restored with an influx of oxygen and reversion to oxygen-using respiration, our usual form of metabolism.) Fermenters thrive in environments with little or no oxygen, converting food molecules to useful energy and waste products such as carbon dioxide, alcohols, and acids (including lactic acid). The various distinctive waste products are what confer the wonderful and often quite specific aromas and flavors of fermented foods.

If you would like to sample any of these bacterially fermented foods, beware of large grocery stores, where such items might have been pickled simply by adding acids during the processing. Instead, seek out good delis where pickles, olives, sausages, or sauerkraut are being produced on-site to be sure that you're getting food made using gram-positive bacteria.

TABLE 10.1. Foods and Drinks Enhanced by Gram-Positive Bacteria

Food	Genus of bacteria used			
	Lactobacillus converts sugars to lactic acid	*Pediococcus* produces lactic acid	*Leuconostoc* converts malic acid to lactic acid	*Propionibacterium* makes propionic acid and a cheesy smell
Some bread	√			
Some wine			√	
Hard cider			√	
Some distilled grain alcohol			√	
Some pickles	√		√	
Sauerkraut and kimchi	√	√	√	
Fermented beans	√	√		
Some sausages	√		√	
Some olives	√			√

Note: Table excludes cheeses, which are too diverse to be summarized here.

Cheeses

The first humans to look at a crock of coagulating, spoiled milk were surely hungry as they took their initial tastes. It is fortunate for all of us that, over the ages, humans from different cultures were adventurous enough to sample all sorts of dairy products that had "gone by" or, as the writer Clifton Fadiman said of cheese, made a "leap toward immortality." And we are particularly indebted to the French, who experimented hundreds of times to produce some of our most exquisite examples of rotten milk, or cheese. Interestingly, cheese did not become part of the cuisine of China or the Americas, though these cultures did develop other fermented foods, such as bean curds and fish sauces, using various gram-positive bacteria. Genetic differences in the abilities of adult humans to digest milk may have influenced these cuisines (although many people with an intolerance for lactose can tolerate lactic acid in fermented milk products).

If you are nibbling on a bit of cheese, you are most likely enjoying the by-product of gram-positive bacteria. The cheese itself is their field mark. Most cheeses are produced by the action of several types of gram-positive bacteria. Details concerning specific culture methods and strains of bacteria can be found in manuals and guidebooks on cheeses and cheese making. Many of the commonly used gram-positive bacteria may be purchased freeze-dried from supply catalogues for cheese makers. A few cheese makers, however, bypass bacteria and use an acidifier such as lemon juice or an enzyme such as rennet to acidify and coagulate the milk.

One of the useful things about cheese or any bacterially enhanced food is that the presence of discrete bacteria and their products often prevents other less desirable bacteria from growing in the same place. Bacteria and their wastes (especially acidic wastes) actually work as natural food preservatives, allowing cheeses and some fermented foods to age well. Certain good flavors and aromas even become enhanced over time. So, although a pail of fresh milk is highly perishable (even today, with a refrigerator), many cheeses may be stored for long periods and enjoyed when needed. Ironically, pasteurization (heating to 63–75°C [145–167°F]) to kill microorganisms, which is a common preservative method for dairy products, makes it *more* likely that the invading bacterial contaminants will not be of the pleasant cheese-making variety. Thus, many recipes for making cheese begin with raw milk—spoiled pasteurized milk is unlikely to produce edible cheese.

FIELD MARKS OF GRAM-POSITIVES OF CHEESES
Note that some of the bacteria listed here make appearances in more than one product. Also, other microorganisms, especially fungi, make their own important contributions to the production of cheese, but these are not treated in any detail here.

Propionibacterium
Propionibacterium ferments lactate, a major ingredient of milk, into propionic acid, acetic acid, and carbon dioxide gas. Cheeses such as Swiss cheese with large "eyes" (formed as bubbles of carbon dioxide) are a result in part of the activity of propionibacteria. Propionic acid has been described as smelling rancid, pungent, and disagreeable. Acetic acid is vinegary. Somehow in small quantities, however, these two acids give a distinctive and delicious taste to the large-eyed cheeses. Propionibacteria are also associated with the sebaceous glands of skin (see chapter 12).

Brevibacterium
Brevibacterium appears as an orange-red covering on some strongly flavored surface-ripened (rind-washed) cheeses (plate 47), such as Limburger, Liederkrantz, Bel Paese, Pont l'Evêque, Port du Salut, Muenster, and sometimes Stilton, which is also full of the blue fungus penicillin. Well-aged Camembert and Brie made by traditional methods may also acquire a coating of brevibacteria. Note that a white, feltlike covering found on Camembert or Brie is white penicillin, a fungus. Be sure that the red color on the outside of the cheese isn't just dye, as it is in American versions of Muenster.

Brevibacterium is also featured in chapter 12 as a cause of foot odor, because it tends to break down proteins from flaking skin between the toes to form a smelly sulfur compound. Is there a connection between Limburger and foot odor? Perhaps. Smell them for yourself; the same sort of bacterial process of breaking down protein occurs in both. Some of these cheeses do smell quite strongly of dirty socks. In small quantities, that smelly waste product of *Brevibacterium* is appealing, at least to some well-trained palates.

To assist me in the research for this book, friends provided interesting cheeses with bacterial pedigrees. One friend sent a very ripe wheel of Muenster cheese to me by mail, wrapped in several layers of plastic. During transit, it became so ripe that the post office wrapped the box in additional plastic. Nevertheless, the package reeked. At first, before I began unwrapping, it smelled like a leak from a gas stove. Increasingly, however, as I removed the five layers of plastic bags and two layers of plastic wrap, it began to smell like a full diaper pail. I had to take the cheese out of doors to complete its unveiling. I then telephoned my friend, who dared me to eat some. I had two crackers full. The flavor was surprisingly mellow and creamy, albeit quite complex and aggressive in its afterflavors, which lingered in my mouth and digestive system for hours. Delicious as it was, I rewrapped it tightly and discarded it in a public Dumpster. Hours later, the house still smelled of it. Days later, the kitchen wastebasket continued to emit an odor; we discovered a completely dry outer wrapping of the cheese that had retained the smell and transferred it to everything else in the trash.

Surface-ripened cheeses have inspired the essays of many food writers, including Peter Mayle (*French Lessons*) and Jeffrey Steingarten (*The Man Who Ate Everything* and food columns in *Vogue*). Steingarten laments that some of the smelliest (and most delicious) cheeses are forbidden entry into the United States. He warns of the growing menace of FDA Code of Federal Regulations Title 7, Section 58:439, which attempts to protect Americans from overly ripe cheese. Nevertheless, excellent cheese shops in the United States still manage to sell delicious contraband products—including Reblochon, Livarot, and Brie de Melun—which Steingarten describes as "sensory, hedonic, aesthetic, cultured, and spiritual." He explains that the distinctive aroma of such cheeses is due to S-methyl thiopropionate, also called "les pieds de Dieu" (God's feet) by French poet Léon-Paul Fargue. An "unripened" white streak of cheese in the center of an otherwise reeking round is called the "âme," or soul of the cheese.

Lactococcus

Lactococcus contributes its waste products, including lactic acid, to the flavors and aromas of eyeless or small-eyed cheeses, which include Cheddar, Camembert, Tilsit, cottage cheese, and Gouda. Two sour milk products, kefir and buttermilk, are made with the help of *Lactococcus*.

Lactobacillus

Lactobacillus is a famous genus of gram-positives featured in the production of Emmental, Gruyère, Gorgonzola, mozzarella, provolone, and Cheddar cheese, as well as kefir and yogurt. Lactic acid, the waste product of these bacteria, confers a pleasant sour taste and acts as an inhibitor of other microorganisms. (Note that the blue of Gorgonzola is due to blue penicillin fungi.)

Leuconostoc

Leuconostoc participates in the production of buttermilk, kefir, and some cheeses by converting malic acids to lactic acids.

VIEWING GRAM-POSITIVES OF CHEESES UNDER A MICROSCOPE

You may want to try to view some of the tiny round or rod-shaped bacteria that make cheese. Keep in mind that a mature Cheddar, Swiss, or other aged cheese really represents the past actions of bacteria, so these are not good samples for viewing. Instead, try a freshly made soft cheese from a deli, plain yogurt that contains active bacteria, or cheese or yogurt that you have made yourself using cultures of appropriate bacteria. Milk products can, however, present a challenge to microscopists; globules of fat are round and can be mistaken for bacteria. Furthermore, the dense, protein-rich nature of milk may reflect your microscope light in ways that will make seeing the bacteria difficult.

CULTURING GRAM-POSITIVES OF CHEESES

You can learn to make yogurt or one of the soft cheeses that require bacterial cultures. However, this field guide is not the place to begin. Books on cheese and yogurt making will tell you exactly what to do and how to purchase or "borrow" the lactic acid bacteria you need.

Cultured Butter

Upon first consideration, butter may not seem a likely candidate for a bacterial product. The starting ingredient is fresh cream, churned or shaken until it forms butter, which is then rinsed, stored at cool temperatures, and eaten soon after it is made. In short, butter seems too fresh and perishable. There would not seem to be many opportunities for bacteria to participate in the process. However, butter has a long complex history, as described by Margaret Visser in *Much Depends on Dinner*, and traditional forms allow more room for bacterial action. Fresh, salted butter, typical of U.S. supermarkets, is a different product from the type of butter that was developed well before sterile techniques and refrigeration were available. These traditional butters start with cream that has ripened or fermented (via lactic acid bacteria) for several days, resulting in a strong but unsalty lactic acid flavor. This is sometimes called "cultured butter" and is more readily found in Europe or as a gourmet item in the United States. Sometimes manufacturers of fresh, salty American-style butter add a little lactic acid to mimic the flavor that bacteria might produce, although the overall effect is still very mild. Even manufacturers of margarine, striving to reproduce all of the tastes, smells, and textures of butter, add a little lactic acid to their complex nondairy products.

The ancient peoples of Iceland, Finland, and Scotland may have taken the bacterial fermentation of butter an extra step. Apparently some of them buried wooden firkins of butter in bogs for months or even years to improve the flavor. The product seems to have been hard, grayish (or sometimes reddish), and acidic, perhaps closer to being a sort of cheese.

Beers and Ales

Occasionally our human ancestors must have been dismayed to find that their dry stores of grain had become moist and sprouted or, even worse, showed signs of musty fungal and bacterial activities. But hungry humans were probably not too quick to toss out that grain, and they would have discovered that cooking up the mess resulted in a reasonably palatable sprouted-grain porridge, slightly sweetened by the sugars produced by germination and perhaps interestingly flavored by microbial byproducts. Most likely, porridge-like concoctions (rather than breads) were the earliest products of wild grains.

However, what if a large store of grain had begun to sprout and decay, and the resulting vat of porridge was too large a quantity to eat right away? A new set of microbes would certainly settle in as soon as the porridge had cooled. Some of our adventurous ancestors probably tasted the mix after it had been bubbling and frothing for a few days. Fermenting yeast would have been producing ethanol (grain alcohol), and various bacteria would have been adding acidity and complexity to the brew. Thus, beer was born—most likely at many different times around the world, with different types of grains and assemblages of microorganisms. The presence of all those microbes as well as the alcohol would have served as a preservative, keeping the beverage from further contamination by less desirable organisms.

Millennia later, five major types of beer and several minor types are well established as beverages for modern humans. The beers may be roughly divided according to how much bacterial activity is allowed by the brewer. Lagers, porters, stouts, and most ales are the result of yeast (fungal) fermentation, with all bacteria kept to a minimum or, in modern, sterile facilities, excluded completely. These beers feature the bitter taste of hops, an herb, that is used to impart an interesting flavor and also to serve as an anti-microbial preservative. Lagers are brewed at cool temperatures, which also helps to reduce microbial activity.

Brewing seems to have evolved differently in Belgium. Belgian beers and to a lesser extent wheat beers from Berlin (Berliner Weisse) are brewed to enhance the activities of gram-positive bacteria. Belgian lambic styles (including Lambic, Gueuze Faro, and fruit beer), Belgian wheat beer, and some Belgian ales (e.g., reds) are relatively acidic and are full of complex flavors due to a host of gram-positive bacteria. The brewing process welcomes bacteria: windows and sometimes roofs of traditional breweries are kept open so that native bacteria can fall into the vats; cobwebs and dust (presumably full of bacteria) are allowed to adorn the walls; and aged barrels (well inoculated with bacteria) are used to store the beer, sometimes for years, as the complex flavors evolve (figs. 10.1a, b). The Belgian style of brewing may well represent the ancestral method, dating

a

b

from a time when it was not easy to control bacterial activity. Rather than trying to fight the activity of gram-positive bacteria, the brewing process developed to enhance their activities. Jean-Xavier Guinard, author of *Lambic*, cites evidence that lambic resembles sikarn, perhaps the first beer brewed by Sumerians in Mesopotamia, 5,000 years ago.

Belgian-style beers are so complex that home brewing them is a favorite project for some professional microbiologists, who enjoy trying to orchestrate the proper succession of microorganisms. For non-Belgian home brewers, the end result (according to purists) is a "p-lambic" (the *p* is for pseudo) because it is thought that only in Belgium does the correct assemblage of microbes occur naturally in the vicinity of the breweries. Beware of home brewing recipes that suggest simply dumping in some lactic acid to achieve a Belgian-type brew.

Beer drinkers who are accustomed to bitter lagers, stouts, and ales sometimes find Belgian-style brews to be an acquired taste (and vice versa). The panoply of bacteria and the long aging process of the Belgians result in a surprising complexity of taste, with layers of different flavors in the fore, middle, and after tastes.

FIELD MARKS OF GRAM-POSITIVES OF BEERS AND ALES

The Belgian-style beers themselves are the field marks for the following (mostly) gram-positive bacteria (table 10.2):

- *Lactobacillus*: These rod-shaped bacteria are nearly ubiquitous in fermenting foods of all kinds. They convert sugars to lactic acid and give a pleasant acidic flavor to some beers. Brewers of lambics, Belgian wheats, Belgian red ales, and Berliner Weisse either add them deliberately to the brew or allow them to tumble in naturally in open-air brew processes.

< **FIGURE 10.1a, b.** Traditional Belgian brewing practices include not being overly fastidious about cobwebs and dust—the source of a diversity of gram-positive bacteria and "wild" yeast that provide a panoply of flavors and aromas in Belgian beers. Here, windows allow access to a fermentation vat (a), and well-aged wooden barrels host a community of bacteria (b). (Photographs by Randy Thiel.)

TABLE 10.2. Gram-Positive Bacteria Used in Brewing Beer

Beer type	Bacteria used	Practices or ingredients used to control numbers of bacteria	Flavors, aromas
Lambic style (Belgian) Lambic, Gueuze, Faro, fruit beers	*Lactobacillus* *Pediococcus* *Leuconostoc*	Aged hops	Acidic, fruity, cheesy, woody, toasty, vegetal, earthy, horsey, leathery, barn-like
Wheat beers			
Berliner Weisse, Belgian Wheat	*Lactobacillus*	Mild hops	Refreshing, lemony, acidic,
Weizen beer	None	Mild hops	flowery, sparkling, fruity, spicy
Ales			
British style	None	Strong hops	Usually bitter and malty, smooth, fruity, sometimes chocolaty
Belgian reds	*Lactobacillus* *Acetobacter*	Mild hops	Sweet and sour, tart
Porters and stouts	None	Strong hops	Roasted, toasty, coffee- or cocoa-like, a little fruity, rich
Lagers	None	Brewing in the cold (~0°C / ~32°F) Hops	Clear, simple, refreshing, crisp

Source: Michael Jackson, *Michael Jackson's Beer Comparison* (Philadelphia: Running Press, 1993).

- *Pediococcus*: This little round bacterium is one of the many lactic acid producers that is prevalent in food. Brewers of lambic add it deliberately or allow it to fall into their vats. Other brewers who use sterile, modern methods consider it to be a contaminant.

- *Leuconostoc*: These oval bacteria produce lactic acid from malic acid, which is found commonly in fruits. Some lambics containing fruit are made in part by *Leuconostoc*, which mellows the sharp fruit acids.
- Acetic acid bacteria: These bacteria are gram-negatives, described further in chapter 6. However, they are mentioned here because as you sip your ale, one of the many flavors you detect might be due to acetic acid bacteria. Belgian red ale has a surprising sweet and sour flavor, which is partly the result of bacteria that produce acetic acid from sugars.

VIEWING GRAM-POSITIVES OF BEERS AND ALES UNDER A MICROSCOPE

Purchase some Belgian-style beer, especially some that has a bit of sediment in the bottom. If it has not been pasteurized, you may be able to detect tiny rods, cocci, and round to oval yeast cells. If you are an advanced home brewer and have access to a good supply of brewing cultures, you can make your own Belgian-style beer and observe microbes at almost any time during the brewing process. There is at least an order of magnitude difference in size between yeast (fungal) cells and bacterial cells. (The contrast is similar to that seen in plate 69 depicting the size difference between algae and cyanobacteria.)

CULTURING GRAM-POSITIVES OF BEERS AND ALES

First learn to make ordinary beer, which is really quite a simple matter. Stores that sell the ingredients for brewing also have manuals on how to brew your own beer. Once you are feeling confident with the process, attempt one of the Belgian styles using bacteria as well as a yeast. Be prepared to experience the growth and development of a complex microbial community. Over the course of months (and even years), your vat will develop floating bacterial rafts and even "ropy" strings of microbes. The process is so complex that you may want to talk to other lambic brewers to assure yourself that everything is going along fine. Let your first batch age well, and then taste with caution. An unconventional guide to brewing is *Wines and Beers of Old New England* by Sanborn Brown, which includes many matter-of-fact descriptions of bacterial fermentation and explicit instructions for encouraging it. This book also includes directions for making "spit malt," in which the source of bacteria is your mouth. Some traditional corn and other beverages of Central and South America undergo a similar process, being chewed first and then placed in a fermentation vat.

Sourdough and Sour Grains

In the past, fermented grains not destined to become beer were likely to become food. Our best and most delicious modern-day example is sourdough bread, which must have originated when lactic acid bacteria dropped accidentally into dough. These bacteria, along with wild yeast, would have produced leavening (rising caused by bubbles of carbon dioxide) and an interesting sour flavor. Serious bakers keep cultures of lactic acid bacteria and yeast in the refrigerator, occasionally replenishing their food—wheat flour—and encouraging the bacteria to grow in warm temperatures overnight before baking day.

Fermented cereals and porridges are more unusual foods today, although they must have been common, and in fact unavoidable, foods for our ancestors. Few recipes survive. No matter how fashionable it might be to try to reproduce peasant cookery in today's kitchens, we generally draw the line at leaving food at room temperature for a few days to acquire a floc of bacteria. However, here is a recipe for Welsh or Scotch surghans, or sowans (oatmeal jelly), from *The Old World Kitchen* by Elisabeth Luard:

> Steep the inner husks of threshed oat grains in water for 4 to 5 days, pour through a sieve, then allow to stand 2 more days. Pour off the liquid, then cook up the residue as a sour porridge. Enjoy.

You can bet that bacteria were involved in producing that sour flavor over the course of the 6- or 7-day incubation.

The Russian beverage kvass is an interesting variation on the sour grain theme (fig. 10.2a, b). It is a beerlike drink produced by the fermentation of bread, especially Russian rye bread, which is often made using a sourdough process.

a

b

FIGURE 10.2a, b. The author sipping kvass from a vendor in Moscow (a); the vendor's truck (b). Kvass is a fermented, beer-like drink made from rye bread.

A tiny sample of sourdough starter under the microscope is likely to reveal little spheres and rods of lactic acid bacteria along with large yeast cells.

You can learn to make sourdough bread using starter cultures of bacteria from a natural foods store or baking supply source. For a more traditional approach, try a longer process in which you allow lactic acid bacteria from the air to fall into your dough and induce fermentation. Recipes are given in many cookbooks.

MALOLACTIC WINES

Basic winemaking usually involves the controlled fermentation of fruits, especially grapes, by yeast (fungi), which produce ethanol as well as acids and other compounds that result in a complex alcoholic beverage. Most modern winemaking techniques conscientiously avoid "wild fermentation," which might include not only inappropriate local yeast but also various bacteria, usually gram-positives. This is one of the reasons that sulfite (SO_2) is so commonly added to wine as an antibacterial and preservative.

In some cases, however, bacteria are encouraged to participate in the fermentation process, with interesting results. Notable among these bacteria are gram-positives such as *Leuconostoc*, which ferments malic acid to the smoother, more mellow lactic acid. Some of these bacteria also give lambics their complex flavors and aromas. For certain red wines that are expected to become more complex as they age, such as premium Burgundy wines, malolactic fermentation is used in the winemaking process. It is a sort of second fermentation that occurs after the yeast have completed their work and all the sugar is gone. Customarily, malolactic fermentation occurred in barrels of red wine during the springtime, when sap was rising in the grapevines of the vineyard. The extra malolactic fermentation

PLATE 1. Porcelain Basin in Norris Geyser Basin of Yellowstone Park. The brownish colors are bacterial pigments, and the white deposits are the mineral sinter, or geyserite. (Chap. 1)

PLATE 2. Morning Glory Pool in Upper Geyser Basin of Yellowstone. The deep blue center indicates a temperature near boiling. White geyserite forms the first ring and indicates a temperature greater than 75°C (167°F)—an ideal habitat for hyperthermophiles. The yellowish ring, less than 75°C, contains photosynthesizers, most likely green nonsulfur bacteria. (Chap. 1)

PLATE 3. Runoff from Excelsior Geyser in the Midway Geyser Basin of Yellowstone. The orange color indicates temperatures of 60–80°C (140–176°F), where green nonsulfur photosynthesizers are thriving. (Chap. 1)

PLATE 4. Grand Prismatic Spring in Midway Geyser Basin of Yellowstone. This spring clearly shows the concentric circles of color described in plate 2. (Photo by Michael Gousie.) (Chap. 1)

PLATE 5. The author demonstrates the convenience of the boardwalk system of Yellowstone Park: you can get quite close to thermophilic bacteria of all sorts. Note: never sample from areas such as this. The bacterial community will show marks for months if disturbed; the U.S. Park Service rightly considers it to be vandalism. In this scene of runoff waters from Punch Bowl, a boiling feature in Upper Geyser Basin, the orange is a mat of green nonsulfurs. (Chap. 1)

PLATE 6. At this unnamed boiling pool at Black Sand Geyser Basin, a beige or flesh-colored layer (green nonsulfurs) lies over a green layer (cyanobacteria). (Chap. 1)

PLATE 7. Methane bubbling up from swamp sediments. Bubbles may appear spontaneously, or you can stir the sediment with a stick. If there is a sulfide smell or a green scum of photosynthesizers, the bubbles may be hydrogen sulfide or oxygen rather than methane. (Chap. 2)

PLATE 8. You can use an apparatus like this to collect methane from a swamp and observe its flammability. (Chap. 2)

PLATE 9. The popping or crackling sounds of water lilies, shown here, and other aquatic plants of shallow, stagnant waters can be considered auditory field marks of methanogens. (Chap. 2)

PLATE 10. "Sulfur Cauldron" at the Mud Volcano area of Yellowstone has a pH of 1–2. The yellow is due to sulfur particles to which *Sulfolobus* and other archaea adhere. (Chap. 3)

PLATE 11. Sour Lake (Mud Volcano Area, Yellowstone) also harbors *Sulfolobus* and related bacteria. The bright green eukaryotic alga *Cyanidium* thrives in the cooler runoff. (Chap. 3)

PLATE 12. Runoff near the acidic Whirlygig Geyser in Norris Geyser Basin, Yellowstone. The bright green is *Cyanidium*, an acid-tolerant eukaryote. The red is oxidized iron, an indication of iron-oxidizing bacteria. (Chap. 3)

PLATE 13. Echinacea Geyser in Norris Geyser Basin, Yellowstone, leaves reddish-brown iron-rich deposits. When active, this acidic geyser is a habitat for iron-oxidizing bacteria. (Chap. 3)

PLATE 14. The pink layer of this salt crust contains halophilic archaea. The black layer is iron-sulfide–rich mud, a field mark of sulfate-reducing bacteria. The slight blue-green color indicates salt-tolerant cyanobacteria. (Chap. 4)

PLATE 15. Thick, black, sulfide-rich mud lies just below the salty surface of this salina in Mexico. (Chap. 4)

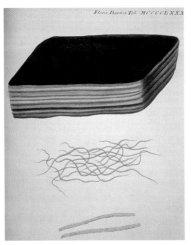

PLATE 16. A plate from the 18th-century work *Flora Danica* depicting what appears to be a very stable sediment, showing years of accumulation of different-colored bacteria. So many distinct layers is uncommon. In an unagitated area of a typical marine mud or salt flat, you might find three or more layers. Filamentous cyanobacteria are shown at the bottom. (Chap. 5)

PLATE 17. A sulfide-rich, relatively stable marine sediment with purple sulfur bacteria on the surface. Lightly brushing away the pink-purple scum reveals a green layer of green sulfur bacteria. A little deeper is a thick, black sediment where sulfate-reducing bacteria are producing hydrogen sulfide. A clump of brown seaweed is at the top. (Chap. 5)

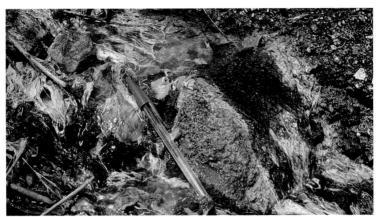

PLATE 18. Green sulfur bacteria growing in a sulfur spring. Filamentous white sulfur oxidizers are shown, too. (Chap. 5)

PLATE 19. A foamy floc of *Zoogloea*. If you look at such a sample under the microscope, expect to see many ciliates, including stalked ones, and worms. Long thin filaments of *Sphaerotilus* may also be present, as well as thicker fungal filaments. (Chap. 6)

PLATE 20. Iron oxide (rust) deposited on plants, rocks, and sediments in the aerated part of a river draining a bog. (Chap. 7)

PLATE 21. Iron slag from an 18th-century furnace in which bog iron was being produced. (Chap. 7)

PLATE 22. Gleyed soil (or clay) with patches of iron-oxide–rich clay. (Chap. 7)

PLATE 23. Silvery-white deposit of manganese and iron oxide at the edge of an iron-rich pond. (Chap. 7)

PLATE 24. Manganese and iron oxide film on the surface of still water near an iron bog. Although it looks like oil, the crust is metallic and breaks up when poked with a stick. (Chap. 7)

PLATE 25. Iron oxide runoff from a thermal spring in the Norris Geyser Basin in Yellowstone. (Chap. 7)

PLATE 26. Iron is being dissolved from this buried chain by the activities of sulfate reducers in the marine sediment, and iron is being oxidized in the photic zone. This oxidation could be an indirect effect of photosynthesizers that produce oxygen; iron-oxidizing bacteria may also be present. (Chap. 7)

PLATE 27. Iron oxide on a marine mollusc shell. (Chap. 7)

PLATE 28. Red iron oxide in a marine environment with underlying sediments of black iron sulfide (pyrite) and and a thin crust of manganese and iron oxides on the surface of the water. (Chap. 7)

PLATE 29. Desert varnish, a manganese oxide stain found on desert rocks. (Chap. 7)

PLATE 30. A procedure for enriching iron oxidizers. (Chap. 7)

PLATE 31. A film of manganese and iron oxide on the surface of the modified Winogradsky column. (Chap. 7)

PLATE 32. Buried iron object from a bog. The flag allows for easy retrieval. (Chap. 7)

PLATE 33. Use a garden soil test kit in any water or soil environment. Here, the test shows that iron-rich bog water is acidic (tube on left), and very low in nitrogen and potassium (second and fourth tubes from left), but high in phosphate (third from left) because of the lack of organisms in the phosphate cycle (phosphate remains bound to iron). (Chap. 7)

PLATE 34. The rounded, pitted contours of this grave sculpture of a reclining lamb indicate the activity of ammonia oxidizers as well as lichens. (Chap. 7)

PLATE 35. Myxobacteria (tiny black dots) growing on rabbit dung with undigested plant material (Chap. 8).

PLATE 36. Shells of two molluscs with adaptations for life in a sulfuretum. The razor clam (*Ensis*; left) moves up and down in sediments according to changes in oxygen levels. The lucinid (right) has symbiotic bacteria that provide nutrition for the mollusc. (Chap. 9)

PLATE 37. The tube of the plumed worm *Diopatra* may be black with sulfide in some environments; microscopic worms (some of which have bacterial symbionts) live on the tube. (Chap. 9)

PLATE 38. *Beggiatoa* (sulfide oxidizers) forming a white film on a marine intertidal sediment. Black iron sulfide lies directly beneath. (Chap. 9)

PLATE 39. *Beggiatoa* forming a white film on green seaweed. These fast-moving bacteria optimize their location in interfaces between sulfide-rich waters (or sediments) and oxygen-rich waters. Here, the bacteria may be attracted to the oxygen production of the seaweed in the otherwise sulfury water. (Chap. 9)

PLATE 40. This outlet from a warm sulfur spring shows cyanobacteria forming two black bands flanking purple sulfur bacteria and white sulfur oxidizers in the center. (Chap. 9)

PLATE 41. A bloom of purple sulfur bacteria on an intertidal flat; the greenish color of the emerged section is cyanobacteria. As the tide changes, the purple sulfurs move to areas where sulfides are concentrated. (Chap. 9)

PLATE 42. A closer view of the purple sulfur bacteria shown in plate 44. Patches of white *Beggiatoa* can also be seen. (Chap. 9)

PLATE 43. A Winogradsky column provides an excellent way to view the zonation of bacteria in a sulfur-rich environment. The top layer is composed of cyanobacteria and often eukaryotic green algae as well. Below that is a layer of purple sulfur bacteria, along with white sulfide oxidizers. Sometimes green sulfur bacteria are evident just under the purples, as in this column. Black sediments at the bottom are rich in many bacteria, including sulfate reducers that generate sulfides. (Chap. 9)

PLATE 44. The waters and sediments of alkaline (soda) lakes or springs may be colored pink-purple by bacteria in the family Ectothiorhodospiraceae if sulfides are present. The white areas are precipitated minerals. (Mono Lake, photograph by John Kricher.) (Chap. 9)

PLATE 45. Black iron sulfide (pyrite) along with a sulfide smell is a good indicator of sulfate reducers. Here, the blackened sediments are exposed at the sediment surface along the edge of a mat of purple sulfur and other bacteria. The black sediment continues under the mat. (Chap. 9)

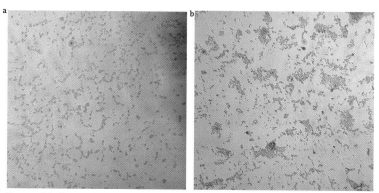

PLATES 46a,b. Gram-positive bacteria (a) retain blue stain, whereas gram-negative bacteria (b) do not. The gram-negative cells have been stained pink; otherwise they would be colorless. (Intro. to Gram-Positives)

PLATE 47. Several types of strong-smelling cheeses with pink-orange-red surfaces are products of brevibacteria. (Chap. 10)

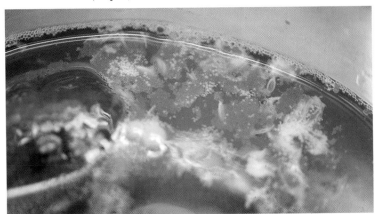

PLATE 48. Scum and bubbles from a crock of fermenting cucumbers—soon to be pickles. (Chap. 10)

PLATE 49. Threads of fungi (not actinomycetes) coursing through a rotting log. (Chap. 11)

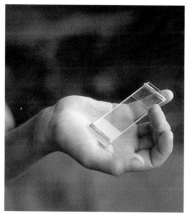

PLATE 50. A peloscope before burial. (Chap. 11)

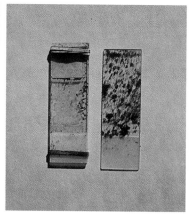

PLATE 51. A peloscope (left) and a plain glass slide (right) after about 10 days of burial in garden soil. (Chap. 11)

PLATE 52. Manatees are rotund, slow-moving marine mammals that use bacterial symbionts to digest algae. (Chap. 12)

PLATE 53. *Acanthurus* (surgeonfish; black fish in photo) digest algae with the help of bacterial symbionts, including *Euplopiscium*, one of the largest known bacteria. (Photo by Steven Oliver.) (Chap. 12)

PLATE 54. Bromeliads often have microbial communities in their water-filled tanks. Bacteria and other organisms may contribute fertilizer-like nutrients to the plants. (Chap. 12)

PLATE 55. Pitcher plants are "carnivorous" in that they supplement their nitrogen-poor diet by digesting drowned insects. Bacteria assist in the digestion of their "prey." (Chap. 12)

PLATE 56. Bright green scum, shown here growing in a pond, is not cyanobacteria but rather green, eukaryotic algae. Rule of thumb: if it is too green, it isn't cyanobacteria. (Chap. 13)

PLATE 57. The subtle brownish film on these aquatic plants is a good indicator of cyanobacteria. (Chap. 13)

PLATE 58. Large globular colonies of *Nostoc*. (Chap. 13)

PLATE 59. Cyanobacteria are often found growing on or beneath the surfaces of mollusc shells. (Chap. 13)

PLATE 60. Cyanobacteria outlining the crevices of a fossil about the size of a thumbnail in a carbonate rock. (Chap. 13)

PLATE 61. A cohesive felt of *Lyngbya* and other cyanobacteria in the form of a microbial mat. When dry, the felt forms a leathery crust that may blow in the wind. (Chap. 13)

PLATE 62a. A low, flat cyanobacterial mat on a marine sediment. Bubbles of oxygen indicate active photosynthesis. (Chap. 13)

PLATE 62b. Just beneath the cyanobacteria lies a layer of purple sulfur bacteria. (Chap. 13)

PLATE 62c. A low, flat mat of cyanobacteria under mangroves in Florida. (Chap. 13)

PLATE 63. *Farbstreifensandwatt* (color-striped sand bar) is the German term for colored bacterial layers in sediment. Cyanobacteria are just above a layer of purple sulfur bacteria and black sulfate reducers are just below. (Chap. 13)

PLATES 64a–c. Convoluted or blister mats that form when gases from beneath cause cyanobacterial mats to buckle up. (Chap. 13)

PLATE 65a. Polygonal mats, a form that sometimes occurs in dry conditions when the mats shrink apart. The edges sometimes curl, forming "tee pees." (Chap. 13)

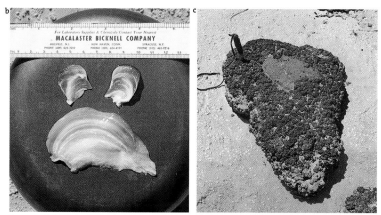

PLATES 65b,c. Sections of a laminated biscuit, showing mineralized layers (b) and pinnacle mats (c). (Photos by Stjepko Golubic.) (Chap. 13)

PLATE 66. A living stromatolite of the type that forms at some sites in the Bahamas. Cyanobacteria faintly color the white carbonate rock. (Photo by Pieter Visscher.) (Chap. 13)

PLATE 67. Typical felt-like filamentous cyanobacterium, *Lyngbya*, as seen with the naked eye. (Chap. 13)

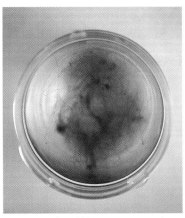

PLATE 68. Thread-like *Spirogyra*, a eukaryotic green alga shown for contrast with plate 67. (Chap. 13)

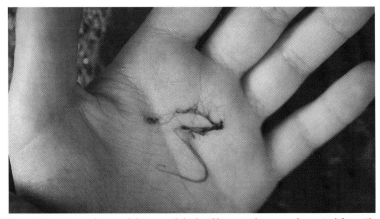

PLATE 69. *Spirogyra* has much larger and thicker filaments than cyanobacteria. (Chap. 13)

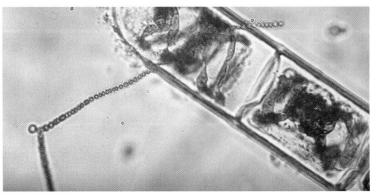

PLATE 70. A mixture of the eukaryotic alga *Spirogyra* and the much smaller cyanobacterium *Nostoc* (on the left), showing the difference in cell size. (Chap. 13)

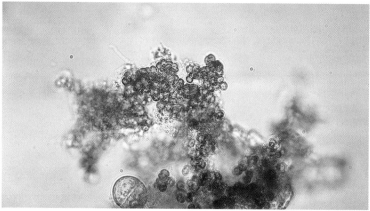

PLATE 71. *Gloeocapsa* under the microscope. Cells are about 3 micrometers in diameter. (Chap. 13)

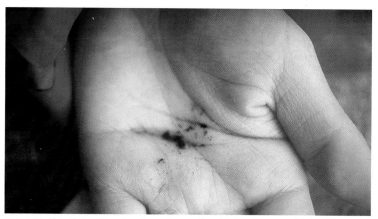

PLATE 72. This photo of *Lyngbya* shows its very fine, felt-like filaments. (Chap. 13)

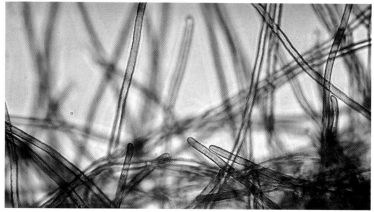

PLATE 73. *Lyngbya* under the microscope. Width of the filament is about 2 micrometers. (Chap. 13)

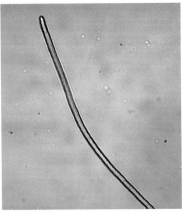

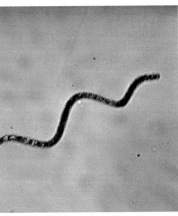

PLATE 74. *Oscillatoria* under the microscope. Width of the filament is about 6 micrometers. (Chap. 13)

PLATE 75. *Spirulina.* Width of the filament is about 4 micrometers. (Chap. 13)

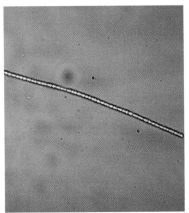

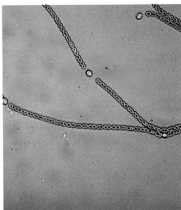

PLATE 76. *Anabaena.* Cells are 4–5 micrometers in diameter. (Chap. 13)

PLATE 77. *Nostoc.* The larger cells are heterocysts, where nitrogen fixation occurs. Cells are 2–3 micrometers in diameter. (Chap. 13)

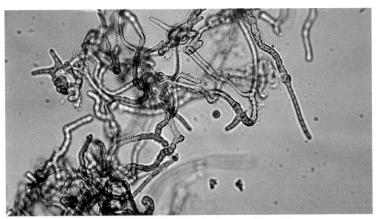

PLATE 78. *Fischerella.* Cells are 2–3 micrometers in diameter. (Chap. 13)

PLATE 79. Cyanobacteria (blue-black pigment) in gypsum that was hammered out from a cliff. (Chap. 13)

PLATE 80. Cyanobacteria beneath the mouth of the gold figure on a fountain at the Summer Palace, near St. Petersburg, Russia. (Chap. 13)

PLATE 81. Cyanobacteria in a protected niche of a gravestone. (Chap. 13)

PLATE 82. Cyanobacteria (moist, black scum) under a shadowy downspout. (Chap. 13)

PLATE 83. Cyanobacteria growing on the damp wall of a carbonate cave. (Chap. 13)

PLATE 84. Crust formed by cyanobacteria on a sand dune. (Chap. 13)

PLATE 85. Cyanobacteria on a dune showing cohesiveness despite the disruption caused by a footprint. (Chap. 13)

PLATE 86. A cyanolichen. Often a blue-green color is indicative of cyanobacterial symbionts. (Chap. 14)

PLATE 87. *Polytrichum*, a moss, growing on a dry, nutrient-poor sand dune with the help of nitrogen-fixing cyanobacteria. (Chap. 14)

PLATE 88. *Bryum*, a moss, growing between sidewalk pavers. (Chap. 14)

PLATE 89. *Azolla*, an aquatic fern, contains cyanobacterial symbionts. (Chap. 14)

PLATE 90. The coralloid roots of a cycad, shown here, contain cyanobacteria. (Chap. 14)

PLATE 91. The yellowish-green color of three-toed sloths is due to cyanobacteria that dwell in special grooves of their fur. (Photo by John Kricher.) (Chap. 14)

PLATE 92. Straw enrichment used to cultivate bacteria able to consume cellulose and other tough plant products. Some bacteria are likely to be in the *Bacteroides* group. Methanogens also are present; they would be dwelling in the anaerobic depths of the jar and producing some of the bubbles. (Chap. 15)

PLATE 93. Octopus Spring, near Great Fountain Geyser in Lower Geyser Basin, Yellowstone National Park. This is where *Thermus* was first discovered, but the area has subsequently cooled and changed configuration. (Chap. 17)

PLATE 94. Cistern Spring in Norris Basin. This silica-rich pool is one of the many sites where *Thermus* is found in Yellowstone Park. (Chap. 17)

PLATE 95. The dusty surface film of a still pond is a good place to seek out organisms of a neuston community, including planctomycetes and other stalked organisms. (Chap. 18)

PLATE 96. Duckweed provides an excellent substrate for attached communities of microbes. (Chap. 18)

PLATE 97. Foam and bubbles below a dam draining an acidic bog. This field mark indicates an abundance of organic materials and microbes; it is essentially the agitated version of the calm surface film shown in plate 95. (Chap. 18)

PLATE 98. Hang a glass microscope slide for hours to days in almost any body of water, and you will quickly accumulate a fascinating attached community of microbes. (Chap. 18)

was said to be "in sympathy" with the rising sap. Malolactic fermentation can be overdone, producing unsubtle effects such as odd flavors and gases. These seem to be tolerated better by drinkers of lambic than by drinkers of fine Burgundy. Decanting can release some of the gas, but too much fermentation can result in complete spoilage.

Some white wines, especially if fermented from grapes grown in cooler climates, benefit from the reduced acidity resulting from malolactic fermentation. These include certain wines of Alsace as well as some North American wines. Other wines are allowed to become gassy or bubbly from bacterial fermentation, producing a wine similar to a mild champagne (in real Champagne, the bubbles are produced by yeast). These slightly fizzy ("frizzante") wines include Portuguese Vinho-Verdes, Prosecco, and certain Italian Piedmonts (white and red) and Northern Italian reds.

CULTURING GRAM-POSITIVES OF MALOLACTIC WINES

If you are an adventurous home winemaker, you might consider branching out with malolactic fermenters. Look for instructions in winemaking manuals.

PICKLED FOODS

CUCUMBERS

The whole idea behind pickling vegetables is to set up a desirable microbial community in and on a vegetable (e.g., cucumber) so that less desirable (rotting and putrifying) communities cannot take hold. It is an ancient idea in food preservation and most likely came about serendipitously. Some bacterial infestations left food inedible, some did not. The latter were cultivated as pickling bacteria. Discoveries of fermented foods include those that resulted from the burial of vegetables (such as cabbage) for winter storage. In some cases gram-positives from the nearby soil must have invaded and converted the cabbage to something like sauerkraut. Thus some traditional peasant recipes for sauerkraut entail burying vegetables in wood- or straw-lined trenches in hopes of attracting and maintaining the right bacteria. This and other extraordinary pickling methods are vividly described by Sue Shephard in *Pickled, Potted, and Canned.*

However, the pickling section of most cookbooks is not likely to mention the word *bacteria*, and modern cookbooks will even try to guide the user away from a bacterial type of pickling process. With growing awareness of bacteria as agents of disease, home economists in the early 20th century developed more aseptic pickling methods, which are usually promoted today. But good, old-fashioned delis are likely to still be making their own delicious and safe fermented pickles, and some comprehensive cookbooks include instructions (as well as warnings) on how to set up a traditional pickling mixture. The method is often referred to as a "long brine" or "old-fashioned brine" procedure. (Avoid recipes that use a typical "canning" or "sterilization" bath; you do not want to kill your bacteria!)

Typically, the cucumbers are first placed in a salt solution in a crock or large jar, sometimes with added vinegar and almost always with a selection of herbs and spices. (Note that vinegar is courtesy of the acetic acid bacteria described in chapter 6.) The cucumbers are held below the liquid under a weighted plate. The crock is left in a cool place for 2 to 5 weeks (plate 48). If the recipe does not take weeks, it isn't a bacterial fermentation. It is a good idea to monitor the fermentation process by removing scum that forms on the surface and in general checking for unappetizing smells and textures, such as slime. If at the end of the procedure, the cucumbers smell, feel, and look rotten, then they most likely are. You made a mistake; don't eat them! However, usually this procedure works, producing delicious results.

KIMCHI

Kimchi, sometimes called the national food of Korea, is a complex pickle of Chinese cabbage along with other vegetables such as radish, pepper, onion, and garlic. Spices and fermented seafood are typically added. A large assemblage of gram-positive bacteria participate in producing kimchi; these include *Lactobacillus, Streptococcus, Leuconostoc, Pediococcus,* and many other types. Kimchi is so important that Korean households often have their own "kimchi refrigerators"—specially designed fermentation boxes that keep the fermentation at the cool soil temperature favored by the microbes. Thus, a steady supply of kimchi is available to accompany nearly every meal. Kimchi has been described as sweet and sour with a little carbonation and other complex and subtle flavors and aromas.

FERMENTED FISH

There are many versions of fermented fish sauces throughout Asia. Some, such as the Thai and Vietnamese fish sauces, are available in the West. A variety of gram-positive bacteria collaborate in producing the unique aromas and flavors of those condiments. If a fish sauce is salty, halobacteria (see chapter 4) are also likely participants.

Garum is the fermented fish sauce of ancient Rome. According to what is known of Roman recipes, the flavor of garum was central to most dishes—so much so that it would be difficult to reproduce a Roman banquet without it. (If you are attempting to do so, perhaps one of the Asian sauces might be a substitute.)

Many proprietary bottled sauces in typical Western supermarkets are the products of fermentation with fish. These include Worcestershire sauce, a concoction of anchovies, molasses, and chiles, as well as the original un-Americanized form of ketchup, which had its origins as a spicy Asian fish sauce.

The traditional cuisines of Europe include many versions of preserved fish. Some of these, such as pickled herring, are prepared using pickling methods from which it would be difficult to exclude bacteria. Other methods—for example, that used to prepare surstromming, a pungent fermented fish of Sweden, come perilously close to putrefaction (see chapter 4 on halobacteria). In general, whole fish ferment readily or even rot in the refrigerator because their digestive bacteria are accustomed to functioning at "cold-blooded" temperatures. Certain traditional methods of pickling fish essentially encourage the bacteria to "digest" the fish from the inside out. Some readers might be wondering about lutefisk, a specialty of Norway. As far as I can determine, bacteria are (surprisingly) not major participants in producing its odor. Rather, the process of making lutefisk seems to be a reasonably successful attempt to control bacteria through drying and treatment with lye—although, in fact, bacteria-like exudates are nevertheless produced.

FERMENTED BEANS

Famous fermented bean sauces of Asian cuisine include soy sauce (a product of *Pediococcus, Lactobacillus,* and fungi) and miso (a product of *Pediococcus* and fungi). Many more sauces exist, producing an array of complex flavors and aromas. Explore the condiment section of an Asian market and try some of the bottled sauces. Note that, in general, bacteria are not advertised as an ingredient in most foods and drinks. Be assured, however, that any product made of beans that appears to lack the typical blandness of most legumes was made with the help of fermenting microbes.

VIEWING GRAM-POSITIVES OF PICKLED FOODS UNDER A MICROSCOPE

Freshly made pickled vegetables, fish, or beans are likely sources of samples for microscopy and will be full of rod and coccoid bacteria.

CULTURING GRAM-POSITIVES OF PICKLED FOODS

Your best bet is to learn to make pickled vegetables, fish, or beans using traditional recipes from cookbooks or from older relatives and friends. Many procedures are still passed down by word of mouth; you might be doing the world a favor by recording them.

OTHER PRODUCTS OF GRAM-POSITIVES

MONOSODIUM GLUTAMATE

Monosodium glutamate (MSG) is a flavor-enhancing compound that is typically used in Asian cuisine. It is produced from sugar by gram-positive *Corynebacterium.* This is usually done as a large-scale industrial fermentation, rather than in the kitchen. However, corynebacteria may be found in a variety of fermented foods such as cheeses, where they do their part to contribute to the complex flavors.

CHOCOLATE, VANILLA, TEA, AND COFFEE

The term *sweating* is used for the first steps in the production of chocolate and vanilla. First, cocoa or vanilla seeds are fermented in the pod by lactobacteria to enhance the seeds' flavor, color, and aroma. The rest of the process includes roasting and extraction. Coffee beans are released from their surrounding fruit by fermenting bacteria that consume the fruit. Black teas and oolong teas are fermented to bring out certain flavors, whereas green tea is not.

To make a variety of coffee called luwak, the processing of coffee beans is taken far beyond simple "sweating." Produced in Indonesia, luwak beans are fermented in the intestines of the civet cat *Paradoxurus hermaphroditus*—also called the luwak. These animals eat coffee beans and defecate them in a form considered to be enhanced. The flavors and aromas have been described as complex, perhaps rivaling those of the Belgian lambics described earlier. The bean-eating civet cat itself is of interest. The luwak is an omnivore and as such probably has a less extensive community of bacteria in its digestive system than most herbivores do. (For more on intestinal bacteria, see chapter 12.) Apparently, though, the luwak has enough of these bacteria to create interesting coffee beans.

SUMMARY: FIELD MARKS AND HABITATS OF GRAM-POSITIVES OF FOODS AND DRINKS

- cheeses
  - Swiss cheeses and others with large "eyes" (*Propionibacterium* [high GC*])
  - Limburger and others with orange-red coating and strong, distinctive aroma (*Brevibacterium* [high GC])
  - most cheeses and fermented milk products such as yogurt (lactic acid–producing bacteria of several types including *Lactococcus, Lactobacillus,* and *Leuconostoc* [low GC])
- some beers, ales, and wines (lactic acid–producing bacteria of several types including *Lactobacillus, Pediococcus,* and *Leuconostoc* [low GC]) (note that acetic acid–producing alpha proteobacteria [chap. 6] may also be involved)
- sourdough and sour grain foods (various lactic acid–producing bacteria [low GC])
- pickled vegetables (e.g., kimchi) made by traditional methods (various lactic acid–producing bacteria [low GC])

* high concentrations of the nucleic acids guanine and cytosine

- fermented fish products (various lactic acid–producing bacteria [low GC]) (note that in salty products, halophiles [chap. 4] are likely participants as well)
- fermented bean products such as soy sauce (*Pediococcus* and other bacteria [low GC])
- chocolate, vanilla, coffee, tea (various lactic acid–producing bacteria [low GC]) (note that the same types of bacteria are used for curing tobacco; see chap. 11)
- monosodium glutamate (MSG) (*Corynebacterium* [high GC])

Chapter 11

Gram-Positive Bacteria of Soils and Other Substrates

The gram-positive bacteria of soils are abundant and diverse. Experienced microbiologists can find dozens of species in just a few specks of humus. However, with a focus in this field guide on macroscopic field marks, the list narrows down largely to the actinomycetes (*actino*: ray-like; *mycetes*: fungus-like), which have the most reliable indicators. Many form branching thread-like cells, something like the hyphae of fungi but much smaller. Actinomycetes are important members of the high-GC gram-positives (see the introduction to the gram-positives). Indeed, the term *actinomycetes* is sometimes used as an alternative name for the high-GC bacteria, although not all of the latter have thread-like growth. In addition to high-GC gram-positives, this chapter includes two low-GC genera: *Bacillus*, in its capacity as an insect pathogen, and *Lactobacillus*, a converter of natural dyes (also featured in chapter 10, as a food-enhancer).

Other major groups of soil bacteria are proteobacteria that convert gaseous nitrogen to useable forms (nitrogen fixation) (chapter 6) and the cyanobacteria (chapters 13 and 14), many of which also fix nitrogen. Consult the chapters on those groups in conjunction with this chapter to get a more complete picture of soil diversity. Also keep in mind that any unusual qualities of soil, for example, excessive moisture, presence of salt or sulfur, or heat from a nearby thermal spring, should prompt you to seek information in the chapters on swamp-, salt-, sulfur-, or heat-loving bacteria.

GRAM-POSITIVES OF SOIL

Rich soils and compost heaps are absolutely teeming with gram-positive bacteria. In essence, soil is a microbial product; there may be a million or more bacteria per gram of soil, most of them gram-positives. The actinomycetes, which include hundreds of species, make up a large portion of these. The fungus-like threads apparently enable some actinomycetes to bridge from one soil particle to another, or to enclose a soil particle in a netlike fashion. The threads may also be useful in elevating actinomycetes above the soil substrate, thereby allowing spores developing on the tips of the threads to be more easily carried away by wind and water to new places. The spores, characteristic of many of the gram-positives, are tiny, round, and often highly resistant to drying or starvation. These are good adaptations for the vagaries of life in the soil which can go through wet and dry spells that exceed the length of many bacterial life cycles.

If you dig up a trowel full of rich, rotting compost, you may find tiny threads clinging to particles of soil. Alas, for appreciators of bacteria, these are very likely to be fungal, not bacterial, threads (plate 49). Fungi are prodigious converters of humus and compost and are at least ten times bigger than any actinomycete. Although they are invisible, actinomycetes in such a sample of compost are hard at work converting all sorts of compounds, such as decaying organisms and the wastes of organisms, into simpler molecules that may be cycled back to larger organisms. The actinomycetes are an integral part of the carbon cycle.

The distribution of food items in soil may be patchy and transient. If, for example, a leaf falls in a particular spot, the compounds in it will briefly sustain a diversity of microbes, including fungi. Then the leaf will be gone, its component parts already cycling into nearby living plants. It is a competitive world. Many soil fungi and actino-mycetes produce antibiotics, which are presumably used as defense compounds against other bacteria. The diversity of bacteria in soil and their fluctuating abundance make them difficult to character-ize. Many bacteria and fungi are slow, patient growers that do not overpopulate a limited food supply. Others are opportunists, taking advantage of favorable circumstances. One method of determining what is in the soil is to see which bacteria grow on a particular medium. Many bacterial media, however, are richer in sugars and proteins than any trowel full of soil. In the presence of such a medium, these (often dormant) opportunistic bacteria can suddenly grow quickly and abundantly. Similarly, if you dropped a roast beef sandwich with everything on some soil and returned a few days later, you would find it swarming with opportunists of all sizes, including insects. But do these represent the "normal" active biota of ordinary soil? It is unlikely. The normal biota are far more difficult to enumerate and evaluate due to their cryptic habits, slow growth, and subtle activities.

FIELD MARKS AND HABITATS OF ACTINOMYCETES

SOIL

"Geosmin" (ge-: earth; -osma: smell) is the term for the smell of freshly turned soil or compost, or soil that is freshly moistened by rain or irrigation. The air above newly plowed fields is full of geosmin. Gardeners may find it pleasant, as it is a familiar indicator of good soil. Consider geosmin to be a field mark of actino-mycetes, since it is one of the many waste products of these bacteria.

Sometimes geosmin appears inappropriately, such as in drinking water. There, the smell is considered musty and is an indication that the water treatment system contains excessive levels of actino-mycetes. Likewise, musty piles of slowly decaying leaves or over-turned logs may smell strongly of geosmin and dust. The appearance of foam and scum in the waters of a wastewater treatment plant can be caused by floating mats of actinomycetes. Thus, a visit to a treat-ment plant may be a way to view a big, healthy colony of them. (Keep in mind, however, that many other organisms can cause the formation of foam and scum; see chapter 7 for more on bacteria of sewage treatment plants.)

Warmth is another field mark of actinomycetes and other gram-positives, especially in piles of decaying vegetation such as compost heaps. On a cool day, compost heaps can steam with tempera-tures as high as 60 to 65°C (140 to 150°F). A similar phenomenon is the warmth generated by gram-positives in silage or piles of wet hay.

SYMBIONTS

Frankia is a type of symbiotic actinomycete that can be found in special root structures of some plants, called actinorrhizal root nodules. This bacterium is a nitrogen fixer—it takes nitrogen gas from the atmosphere and converts it to ammonia-like compounds that the host plant uses to make proteins.

Relatively few organisms can fix nitrogen, and all of them are bacteria (including some archaea). Whenever nitrogen fixers are found in soil, that soil is enriched with nitrogen in a form that can be used by other organisms such as plants for making proteins. Without nitrogen fixers, the nitrogen in the atmosphere would be completely inaccessible. While many nitrogen fixers are free-living, many others are symbionts with plants. They reside within compartments that are specialized to keep out oxygen, which inhibits the process. To learn more about nitrogen-fixing symbionts see chapter 6 on proteobacteria in root nodules and chapters 13 and 14 on cyanobacteria.

The actinomycete *Frankia* can be found living symbiotically in nodules of *Alnus* (alder trees) and *Myrica* (which includes bayberries and sweet ferns). Both of these plant genera can grow in somewhat marginal soils. Alders and sweet ferns can tolerate waterlogged, swampy soils, and bayberry survives well on seemingly barren sand dunes. Nitrogen can be limited in both of these habitats. Plants of swampy or extremely dry areas have evolved many different mechanisms for scavenging and conserving nitrogen. For example, insect-eating plants such as the venus fly trap thrive in low-nitrogen boggy areas by using the protein of trapped insects. Keeping nodules of symbiotic *Frankia* to fix nitrogen is another such strategy.

The field mark of *Frankia* is the root nodule of the host plant. Identify some sweet fern with the help of a guidebook to plants, and pull up a few roots (fig. 11.1). Tiny, pale, digit-like structures should be visible clustered here and there (figs. 11.2, 11.3). You can do the same with a bayberry plant, but be aware that it is likely to be part of a plant community controlling dune erosion. Don't pull up the whole plant; instead, check a blown-out part of the dune for any exposed bayberry roots. It can be tough to dig up tree roots; if you happen to find an *Alnus* tree freshly knocked over by a storm, you

Figure 11.1. Sweet fern is one of many plants with root nodules that contain nitrogen-fixing actinomycetes.

Figure 11.2. Nodules of a sweet fern root.

FIGURE 11.3. A massive growth of root nodules on a bayberry plant. Bayberry should not be dug up where it is used to control the erosion of sand dunes. This specimen, exposed by wind, was found in a dune blowout.

might be able to get a look at its root nodules, which in an old tree can form rather large clusters.

BIRD AND REPTILE NESTS

Alligator nests are essentially compost heaps that are piled up by the female during nesting season. The warmth of the pile incubates the eggs within, and the acidic products of bacterial and fungal fermentation help weaken the egg shells so that the young alligators can exit more easily. The odor of geosmin in such a nest is a field mark of actinomycetes. The warmth of the nest as well as a successful hatching is a field mark of many decomposing microbes, including gram-positives. Note, however, that it is dangerous to get this close to an alligator nest, as the mother alligator is often lurking nearby; plan to appreciate the phenomenon from a distance.

Megapod birds (Mallee fowls and brush turkeys) of tropical Asia and Australia also build composting nests. ("Megapod" refers to the large feet of these birds, with which they dig up and arrange their piles of decaying vegetation.) The heat of fermentation incubates their eggs. It's possible that other birds use the warmth of decomposition to supplement their own body heat for incubation; loons and other aquatic birds, for example, are known to build large, damp nests of decaying plant materials.

COMPOST HEAPS ENHANCERS

In the "organic" section of a garden supply store or catalogue, you may find packaged enhancers for making compost. Some of these products contain a mixture of dormant gram-positive bacteria—including thermal-tolerant actinomycetes—that, if applied to the compost heap, will spring to life and help decompose the pile. Some products also contain nutrients, which are a source of food for the bacteria. Other products contain nutrients only, on the supposition that the bacteria are present in your soil anyway and just need a little boost. In any case, "bacteria" or "microbes" are likely to be mentioned only in fine print. The product name usually indicates whether it will start, enhance, or accelerate the composting process. Many organic fertilizers or "conditioners" claim to enhance the activity of beneficial bacteria by supplying nutrients. In the case of products for growing legumes, it is often the proteobacterium *Rhizobium* that is the active ingredient (see chapter 6).

SURFACES OF MONUMENTS

Discolorations of bare, dry limestone and marble stones (including those that have been used to construct monuments and buildings) can sometimes be attributed to gram-positive bacteria such as actinomycetes. A green-gray, powdery patina in places that seldom receive moisture is sometimes due in part to the activity of actinomycetes, although fungi, algae, and other microbes may be participants as well.

VIEWING GRAM-POSITIVES OF SOIL UNDER A MICROSCOPE

Compost and soil are so rich in a diversity of bacteria and fungi that it will be nearly impossible for you to distinguish which are actinomycetes or other soil gram-positives. If you are up to the challenge, take a speck of soil and moisten it with water on a slide, being sure to remove any tiny bits of quartz or other rocky material. Place a cover slip on top. Under the microscope, look for bacteria; perhaps you will see nearly transparent rods and coccoids among the soil particles. Some of those are likely to be gram-positives, including actinomycetes. The position of the soil determines what niche the bacteria occupy. If the soil was taken far from plant roots, they are probably soil bacteria. If the soil has been rinsed or shaken off roots, the occupants are rhizosphere bacteria. If you have to wash vigorously to remove the soil from the root surfaces, they are rhizoplane bacteria ("rhizosphere" refers to the part of the soil containing roots; "rhizoplane" is the external surface of roots). If you "surface sterilize" the roots—by dipping them into bleach, rinsing them in sterile water, and then macerating them—whatever bacteria you get were probably intimate associates of those roots.

You may also try a buried slide technique. Take a glass microscope slide and bury it vertically in soil for 1 to 3 weeks. To remove the slide and the surrounding soil, use either a can opened at both ends or a bulb planter. *Gently* break away the soil. Wipe one side of the slide clean. Remove any bits of quartz on the other side, add a little water and a coverslip, and observe under the scope.

A slightly more elaborate method for viewing microbial landscapes in soil is a peloscope, or "mud-viewer," developed in Russia, where soil microbiology got its start. The original method involves burying small bundles of flat capillary tubing in mud or wet soil and letting them be colonized by bacteria over a period of days to months. The flat tubes are then retrieved and observed under the microscope as though they were slides. Flat capillary tubing, however, is not typically available to the amateur naturalist, so another Russian-style peloscope is recommended. Take two glass microscope slides and between them place two thin glass coverslips, one at each end, so that a narrow space between the slides is formed (plate 50). Clamp the slides together with two rubber bands, then bury the slides in mud or moist soil. This arrangement provides quite a large area on which microbes may grow, and the peloscope may be used as a regular (albeit somewhat thick) microscope slide once the rubber bands are removed and the outside of the glass wiped clean. Expect to see both fungi and bacteria colonizing the peloscope (plate 51). Fungi may be the first organisms apparent, with their long, thick, multi-branching hyphae. Focus down to the level of the bacteria, which are at least an order of magnitude smaller than the fungal threads. (To get a feeling for the contrast in size between actinomycetes and fungi, see plate 70 showing a similar contrast between cyanobacteria and algae.)

CULTURING GRAM-POSITIVES OF SOIL

If you have a compost heap, you are already culturing gram-positive soil bacteria (and many other organisms, large and small). Note that the first stage of a proper composting heap involves low temperatures and a host of different microorganisms and macroorganisms. The heat of decomposition raises the temperature up to 65°C (150°F). At this point, mostly gram-positives will be active in the compost, although there will be fungi as well.

An indoor project might include keeping some compost in a terrarium and watching (and smelling) its progress. Scoop in some soil, add a layer of materials to be composted (e.g., vegetable waste, egg shells, and shredded paper), top with more soil, and keep the mix moist. Monitor the decay over a period of weeks to months. Occasionally mix the compost a little and note whether there is any smell of geosmin, an indication of actinomycetes.

A slightly more ambitious project (described in appendix A) is to use a soil agar medium to grow soil bacteria. It is important that the medium not be too rich, or you will get nothing but fast-growing opportunists, not representative of the normal active members of the community of microorganisms. Look on your agar plates for tiny, fuzzy colonies; larger, sometimes colorful, fuzzy colonies are likely to be fungi. Both fungi and actinomycetes have long, thread-like cells with which they form bridges from soil particle to soil particle. In the case of the fuzzy colonies, both the actinomycetes and the fungi are hoisting themselves up into the air, making it more likely that their spores will be carried by the wind or water to colonize new substrates.

BACILLUS AS A NATURAL PESTICIDE

A widely used natural pesticide against a variety of insects is the gram-positive bacterium *Bacillus thuringiensis*—available either as a freeze-dried powder or a liquid suspension. This species causes fatal diseases in caterpillars and grubs but not honeybees. With extensive contact it can cause allergic reactions in humans. So resilient is *B. thuringiensis* to drying that it can be packaged and placed on the shelf of a garden supply store or sold by mail order through garden catalogues. Look for it in the organic insect control section under various brand names. There are also some mosquito and gnat control methods that use *B. thuringiensis*. The species' name or the abbreviation BT may be prominently displayed or may appear only in fine print on the back of the package. Another *Bacillus* species, *B. popillae*, is marketed as a Japanese beetle control called "milky spore."

Although pathogens are not a primary focus of this guide, it is worth mentioning that the resilience that allows *Bacillus thuringiensis* to be packaged and used in powdered form is the same resilience found in other *Bacillus* species, including *B. anthracis*, the cause of anthrax. Unfortunately, this allows *B. anthracis* to be distributed easily by criminals and terrorists via simple means such as mass mailings. In its more natural state, *B. anthracis* is a soil organism capable of waiting out long dry spells. Like many pathogens, however, it is an opportunist, capable of rapidly multiplying when it encounters a host organism such as a human or other animal. Antibiotics generally work against *Bacillus*; however, *B. anthracis* produces a dangerous toxin, and if an infection is allowed to go untreated the amount of toxin can reach lethal levels.

STREPTOMYCES OF ATTINE ANTS

The gram-positive *Streptomyces* is associated with the Attine ants, a large, efficient group of about 200 species that create and tend "gardens" of fungi (along with various bacteria). Some especially advanced species cut leaves to feed their fungal crops. These ants are found in Central and South America, Mexico, and on the border into the southwestern United States. One of the challenges of being an Attine ant is maintaining only desirable types of fungi, not weedy types that might overgrow the colony. The ants have on their bodies white (or pale) crusty patches of *Streptomyces* bacteria, which secrete antibiotics and antifungals that discourage growth of unwanted microbes.

GRAM-POSITIVES OF BLUE DYES

Of all the natural plant dyes, the blues are the most challenging to extract and use. Look at any book on plant dyes to see how relatively easy it is to achieve browns, yellows, and reddish tints as well as some pinks and purples. The list of blue-producing plants is usually limited to indigo (*Baptisia tinctoria*) and woad (*Isatis tinctoria*). (A good indicator of the dye properties of a plant is the species name *tinctoria* or *tinctorum*—as in the Japanese dye plant *Polygonum tinctorum*, another source of indigo blue.) Blue flowers, blueberries, and most lichens do not usually yield blue dyes. One reason is that blue, purple, and true red colors in plants are especially sensitive to pH and oxygen conditions and readily turn different colors in various alkaline or acid solutions. Try boiling some red cabbage in two different pots, one with baking soda (alkaline/high pH) and the other with white vinegar (acid/low pH), and note the color differences. The pH-sensitive (and somewhat unstable) pigment is anthocyanin and is responsible for most blues (and some reds) seen in plants. Producing dyes from indigo and woad requires the assistance of fermenting bacteria—mostly gram-positives—to achieve the desired color.

The traditional process of producing indigo blue dye from indigo or woad began with steeping the chopped plants in boiling water. The resulting liquid was made alkaline with ammonia and then oxidized by vigorous stirring to incorporate air. The ammonia was typically the product of bacterial fermentation of urine that had been added to the vat. Under alkaline, oxidizing conditions, the blue dye indigotin settled to the bottom and was collected and stored as dried, caked powder.

To redissolve the dye for use, a bacterial fermentation was set up in a dye pot placed near a fire for warmth. Urine was added for alkalinity and a little bran for food for a population of fermenting gram-positives. Some of the bacteria would have made conditions even more alkaline by fermenting the urea (in the urine) to ammonia. (Note that many different bacteria can ferment urea; for example, an ammonia smell can be detected in chicken houses where urea-rich manure accumulates and bacteria thrive.) Such a solution would be relatively anoxic (devoid of oxygen) and would allow the dye to go into solution in a nearly colorless form. When cloth was dipped into the solution and then hung in the air to dry, the dye would become blue upon oxidation and would permanently set in the fabric. A similar process is used to extract a rich red color from the woody roots of madder (*Rubia tinctorum*). Like blue, true red (unadulterated with brown, orange, or purple) is difficult to achieve with natural dyes. Unfermented madder yields duller colors of red. Some early methods for processing the dye in madder include adding yogurt (full of lactic acid–producing bacteria) and chewing on the roots, introducing not only saliva but a panoply of oral gram-positive bacteria.

Bacterial fermentation is also used to produce the rich colors of lichen dyes. Lichens (symbiotic associations between fungi and algae or cyanobacteria) are the other source of natural blues (e.g., *Xanthoria parietina*) or reds (e.g., *Umbilicaria pustulata*). In general, lichens that yield richer colors by fermentation are those that pass the bleach test. Remove a bit of lichen and gently scrape off the tough outer covering using a knife blade. Add a drop of bleach; a red color means fermentable dyes are present. As with indigo, the process of producing dyes from lichens involves fermentation by a community of gram-positive bacteria in a tightly closed vessel to which some ammonia has been added. The mixture is stirred periodically over the course of several weeks. Once it is ready to use, it can be adjusted into the blue or red range with a little alkalinity or acidity, respectively. The original dye for litmus paper, which is used to determine pH, was extracted from lichens.

In general, blue (and bright red) materials that you know have been dyed with plant dyes can be considered indirect field marks of gram-positive fermenting bacteria. Besides their usual use in dying fibers, these fermented pigments have also been used in artists' colors, such as some of the subtle blues of Vermeer. Also, the deep blues in the Virgin Mary's robes depicted in medieval and Renaissance art may indicate a mastery of indigo fermentation by dyers' guilds. And the ancient Britons were reported by Caesar to have painted themselves blue with dye made from woad for the purposes of looking more ferocious in battle.

An interesting complement to the use of bacteria in achieving rich dyes is their use in traditional bleaching processes. Buttermilk containing lactic acid–producing gram-positive bacteria is used to soak linens, which are then laid out in the sun to whiten over the course of several days.

You can experiment with fermented dyes yourself; however, the brief instructions just given are not meant to be used as the actual procedure. Find good books on natural dying and look specifically for fermentation methods to get the right timing and reagents for each step. See also chapter 15 for the use of bacteria in linen preparation.

Gram-Positives and Tobacco

A process by which the flavors and aromas of tobacco leaves could be mellowed and enhanced was discovered serendipitously in the 1800s and is now standard practice in curing tobacco. The process was discovered when large quantities of damp tobacco were pressed into containers for transport. On the way, the leaves were partially fermented by gram-positive lactic acid–producing bacteria and were found to be much improved. Modern curing techniques involve first drying tobacco and then "sweating" the leaves under moist, warm conditions to enhance fermentation by bacteria. For sweating as a process for enhancing foods see the section on tea, coffee, vanilla, and chocolates in chapter 10.

Photosynthetic Gram-Positives

The photosynthetic heliobacteria deserve a mention here even though they lack visible field marks, unlike most photosynthesizers. They are a good example of the diversity of the gram-positives. Many types of metabolism that evolved in other lineages of bacteria also evolved in gram-positives. As with other processes, however, gram-positives do photosynthesis a little differently. Heliobacteria have their own unique bacterial chlorophyll g, which absorbs light in the infrared range. Rather than using carbon dioxide as a source of carbon for making sugar, they use simple organic acids such as acetate and pyruvate. Green nonsulfur and purple nonsulfur bacteria do this too (see chapters 1 and 8). Usually photosynthesizers are full of extra internal membranes on which certain photosynthetic reactions take place, but heliobacteria lack these; their membranes are anomalous in general, giving inaccurate results in gram-staining tests.

However, one characteristic of heliobacteria is especially typical of the gram-positives. The heliobacteria have resting spores that can survive for long periods in dry soil or in other extreme conditions. In fact, the spores can even survive boiling temperatures. The typical photosynthetic environment for heliobacteria is moist soil. They are apparently usually out-competed in more aquatic environments, where photosynthesizers are usually found, and thus were overlooked for a long time. An exception seems to be water-logged rice paddy soils. Heliobacteria may be isolated from these soils, but their field marks are usually obscured by other bacteria that inhabit rice paddies.

GRAM-POSITIVE FERMENTERS

Because gram-positive fermenters (as well as other fermenters) produce hydrogen as a waste product, and bacteria in several other groups require hydrogen, the presence of hydrogen users is often a good indicator of the presence of fermenters.

Hydrogen, the lightest element, forms a gas (H_2), most of which was lost from the Earth during its formation and early years. Although hydrogen is the most abundant element in the universe, it is scarce in our biosphere primarily because the Earth's gravitational field is not sufficient to keep it in our atmosphere. The other reason for the scarcity is the abundance of oxygen (20% in our atmosphere), which reacts quickly with hydrogen to produce compounds such as water.

Given these conditions, there is great competition among those microbes that require hydrogen in their metabolism. Where hydrogen is available (such as in deep sediments or in digestive systems), it is quickly scavenged by bacteria. Very little reaches the atmosphere. The list of hydrogen users includes methanogens (archaea), sulfate reducers and chemosynthesizers (proteobacteria), nitrogen fixers (several groups), photosynthesizers (several groups), and others. In some microenvironments, competition is so severe that all hydrogen users cannot be together in the same space. Happily, some organisms generate hydrogen as a waste product; these include many of the gram-positives, which form a sort of base to the ecological hydrogen chain.

A common strategy of hydrogen users is to get close to a fermenter, in some cases by forming a symbiosis. In many other cases, it is sufficient to be in a nutrient-rich environment where gram-positives are taking advantage of a fermentable substrate such as decaying (or digesting) plant material. Gram-positives are difficult to identify in such environments, but their hydrogen-using associates (e.g., photosynthesizers, sulfate reducers, and methanogens) are often highly visible. Therefore, a fairly reliable indirect field mark for fermenters, especially gram-positives, is the presence of hydrogen-using bacteria of all types.

SUMMARY: FIELD MARKS AND HABITATS OF GRAM-POSITIVES OF SOILS AND OTHER SUBSTRATES

- rich soils and compost heaps with geosmin smell and sometimes warmth generated by fermentation* (high GC**)
- root nodules of plants such as alders (*Alnus*) and bayberry and sweet fern (*Myrica*) (symbiont: *Frankia*; high GC)
- nests of some birds (e.g., Mallee fowl, brush turkeys) and reptiles (e.g., alligator) with geosmin smell and warmth* (high GC)
- "natural" pesticide powders or suspensions found at garden-supply centers (*Bacillus*; low GC)
- white crusty patches on Attine ants of Central and South America (symbiont: *Streptomyces*; high GC)
- fermenting herbal dyes and curing tobacco (various lactic acid–producing bacteria; low GC)
- some lime-green to gray powdery discolorations of limestone and marble (actinomycetes of various types; high GC)

* Soil and compost contain many organisms, including fungi; determining exactly which types are present is not an amateur project. The geosmin odor, however, is a good indicator of gram-positive actinomycetes.
** High concentrations of the nucleic acids guanine and cytosine.

CHAPTER 12

Gram-Positive Bacteria as Symbionts of Animals and Plants

> If microbes were capable of emotion, they would celebrate each time an infant was born. At birth, a new potential host emerges from the protected environment of the uterus, providing pristine surfaces and body cavities as sites for microbial colonization.
>
> G. W. TANNOCK, *Normal Microflora*

A newborn infant is somewhat like Krakatoa, the volcanic island that was sterile—free of life—just after its eruption in 1883 but which soon became completely inhabited by a diversity of microorganisms and eventually larger organisms (animals and plants). Both became host to an increasing number and diversity of organisms. On the subject of these organisms, and on bacteria in particular, this book will continue to be relentlessly optimistic. "Bad" bacteria get most of the coverage in the literature; it is easy to find information ranging from popular to technical on "germs." However, it is relatively unusual to view bacteria as integral parts of the biosystem—which is the primary viewpoint of this book. Here, you will find few descriptions of opportunistic bacteria running amok throughout our bodies. For that sort of thing, see the writings of Berton Roueche, de Kruif's *The Microbe Hunters*, and some of the writings of Oliver Sacks.

SKIN

Most of the microbes found commonly on the skin of humans are gram-positive bacteria that settle in across the landscapes of the body, seeking moist, nutrient-rich, protected crevices. Each of us carries about 100 billion of them on the 2 square meters of our integuments; they are our indigenous microbiota. The body is their world, and its terrain is diverse—from the "desert of the forearm" to the "cool woods of the scalp" to the "tropical forest of the armpit," to use Mary Marple's words. For the most part, our skin bacteria are benign, taking advantage of substances discarded as wastes: sweat, oils, and dead skin cells. They become harmful opportunists only rarely, when they gain access to body cavities or when the immune system malfunctions. Generally, the day-to-day existence of skin bacteria goes unnoticed, with the exception of body odor—to which several species eagerly contribute. Skin bacteria may even prevent pathogenic bacteria from taking hold, by occupying all of the habitable spaces.

However, it is not an easy life for our indigenous inhabitants. We are constantly renewing our skin cells. Every day, huge rafts of dead cells flake off, carrying bacteria with them. The remaining bacteria must quickly reestablish themselves on the fresh surfaces. It is like living in a zone of constant earthquakes and avalanches. And to make it even worse (for the bacteria), we wash ourselves; our little inhabitants must stick on tight to avoid disaster. Fortunately (for them), "washing" in the everyday sense is not the sort of scrubbing process done by surgeons about to operate. Table 12.1 shows the number of bacteria removed from skin by a "contact method," such as gentle blotting, versus a "scrub method" using a detergent. Actual

TABLE 12.1. Aerobic Bacteria Removed from Skin by Different Methods

| Skin site | Number removed per square centimeter | |
	Contact method (light blotting)	Scrub method (vigorous scrubbing)
Armpit	106	2,400,000
Forehead	348	200,000
Back	55	310
Forearm	41	100

Source: W. C. Noble and Dorothy A. Somerville, *Microbiology of Human Skin* (London: Saunders, 1974).

washing (e.g., with a washcloth and soap) may fall somewhere in between. Thus, we rearrange our bacteria and make life a bit more challenging for them, but we do not clean our surfaces of them.

FIELD MARKS OF GRAM-POSITIVES OF SKIN

Our skin bacteria work in concert to contribute to our characteristic body odors. Those that feed on skin oils, in particular, release volatile by-products. One of the reasons that distinctive body odors arrive with puberty is that sebaceous glands provide skin bacteria (e.g., *Staphylococcus*) new edible secretions. The particular aromas produced by some of our gram-positive bacteria represent their most identifiable field marks.

Propionibacterium: These bacteria thrive in the ducts of sebaceous glands of both adolescents and adults. The bacteria rejoice when their host reaches adolescence, because then these glands often become clogged with rich oils, making a very pleasant habitat for bacteria. Propionibacteria are associated with acne but do not necessarily cause it. They are also responsible for making the holes in Swiss and other "large-eye" cheeses. Just as they do in making Swiss cheese, these bacteria produce smelly propionic acid and carbon dioxide as waste products of their metabolism. Active sebaceous glands and the resultant oily skin along with some components of body odor are among their field marks.

Brevibacterium: Between the toes, *Brevibacterium* is busy digesting dead skin and converting an amino acid, methionine, to methane thiol, which has the distinctive smell of socks and feet. Interestingly, this same bacterium is called into service to produce some smelly surface-ripened cheeses such as Limburger (definitely an acquired taste, as described in chapter 10). Brevibacteria do especially well if the fungi that produce athlete's foot are also present, because the fungi loosen up skin cells and make them more available to the bacteria. Some mosquito species are attracted to the smells of brevibacterial products and use them to locate their hosts.

Staphylococcus: Many species of *Staphylococcus* are found on skin, residing in habitats that include the sweat glands, the sebaceous glands, the hair follicles, and the waxes of the ear canal. Those that break down skin oils are among the contributors to our overall body odors.

Rarely, *Staphylococcus* can become an opportunistic pathogen, taking excessive advantage of its host's skin products. Symptoms of "staph" infections are varied but may include impetigo, abscesses, boils, and sties. Extreme staph infections in which the organisms are thriving within the body cavity are a serious matter; they afflict some hospitalized patients, such as those with immune system deficiencies.

VIEWING GRAM-POSITIVES OF SKIN UNDER A MICROSCOPE

Take a sample from a damp, secluded part of your body (preferably a spot unencumbered by antibacterial deodorants); try the armpits or between the toes. Prepare a wet mount and view under the highest power possible. You may be able to detect tiny rods and coccoids. Note, however, that according to table 12.1, a good, hard scrubbing is the best way to release large numbers of bacteria. It is difficult to detect mere hundreds of bacteria, so instead try to get thousands by scraping fairly aggressively, perhaps using the flat surface of a toothpick.

An additional and surprising inhabitant of your body may also be visible under the microscope. This is the tiny mite *Demodex folliculorum*, which inhabits hair follicles—especially of the eyebrows—and consumes skin bacteria and various wastes of the skin. However, *Demodex* can be more difficult to detect than skin bacteria, as the mites are nearly transparent and not very numerous.

CULTURING GRAM-POSITIVES OF SKIN

Perhaps it is enough for you that you culture these bacteria every day with your skin, sweat, and oil. If that is not sufficient, see appendix A for instructions on how to make a moderately rich nutrient agar. Use cotton swabs dipped into water that you have boiled and then cooled with a cover on it. (The swabs and boiled water are close enough to sterile technique.) If you want to do some scraping, use toothpicks. Swab parts of your body and then spread the swab around on your agar plate. Allow the plate to develop a few days in a warm, dark place. You are likely to see shiny yellow and white colonies, some of which are probably the gram-positive bacterium *Staphylococcus*. A few fuzzy fungal colonies might appear, but they are most likely opportunists and not the usual inhabitants of your skin. An experiment commonly done for science fairs is to sample the same area both before and after washing (or scrubbing) with various soaps.

BODY CAVITIES

"Body cavity" refers to those body areas that are continuous with the skin surface but which, due to their functions and locations, are very moist and therefore desirable as bacterial habitats. These cavities include the entire digestive tract from mouth to anus, the nasal passages, and the urogenital organs as far in as the bacteria can manage to colonize. (Generally, in a healthy individual, bacteria do not get in as far as the bladder or the uterus.) "Body cavity" does not refer to the truly internal places of the body that contain the heart, lungs, brain, and other organs. These places usually are kept free of bacteria by the vigilant immune system.

Body cavities are outstanding culture chambers of bacteria, many of which are gram-positives. Our indigenous microbial cells far outnumber our own cells by about a factor of ten. It has been estimated that each of us has about 10^{14} bacteria and "only" 10^{13} of our own cells. It is actually quite difficult to determine what life would be like if we were not hauling around 10^{14} extra cells representing hundreds of different microbial species. At considerable expense and effort, it is sometimes possible to establish and maintain bacteria-free laboratory animals such as rats. Many body functions of these bacteria-free animals are different. Metabolic and cardiac rates are lower. Nitrogenous wastes such as urea seem not to be processed as efficiently. More calories may be needed to survive. There may be a lack of fecal and ammonia odors, which are usually typical of captive animals. Some researchers have noted that the meat of such animals is not as flavorful.

It is difficult to conclude from these studies which species of microbe is doing what and exactly how we humans might be benefiting by being so full of bacteria. Something good is going on, but what exactly is not well understood. *Escherichia coli* (a gram-negative proteobacterium; see chapter 6) synthesizes vitamin K in our intestines, and bacteria-free lab rats can become vitamin K deficient. However, the richer, more diverse diet of humans may make that particular source of vitamin K less important. Sometimes humans who are born with extreme malfunctions of the immune system are confined to isolation rooms to keep them away from bacteria. However, the effects of the isolation are not easy to sort out, because the subject is generally not healthy to begin with.

Some research indicates that intestinal bacteria can modulate which human genes are turned on to make gene products. These include genes that code for nutrient absorption and blood vessel growth. Also, a layer of indigenous bacteria may mediate between our immune system and the various "foreign" agents (e.g., food and bacteria) that are introduced into our intestines.

It *is* fairly certain that our particular assemblages of microorganisms help to keep other possibly disruptive bacteria from colonizing. If you've ever taken a course of ampicillin, neomycin, kanomycin, or another powerful antibiotic, you might have experienced terrible intestinal cramps as your normal microbiota were nearly wiped out, allowing the remaining bacteria to proliferate and produce all sorts of inconvenient gases and wastes. Many of the beneficial bacteria of the digestive system are lactobacilli (of cheese fame); eating yogurt (full of lactobacilli) seems to help reestablish conditions so that your own lactobacilli can recolonize. The yogurt lactobacilli, such as *Lactobacillus bulgaricus*, typically are not the same as those found in intestines, such as *Lactobacillus acidophilus*. However, some specially blended yogurts as well as acidophilus milk do contain *L. acidophilus* and are presumably even better for recolonization—check the ingredient list to see what bacteria are included. Freeze-dried cultures of *L. acidophilus* may be purchased over the counter at some drugstores for the purpose of maintaining or restoring a proper balance of good bacteria. This type of treatment has been called "probiotic" rather than "antibiotic." Eating yogurt may also be a solution to the problem of bacterial growth on throat prostheses that are used in the treatment of diseases such as throat cancer. Furthermore, *Lactobacillus* when applied to wounds has been found to prevent the growth of pathogens. Similar strategies using a special strain of *Streptococcus* can also be used to control ear infections.

Large-scale operations for raising animals for meat are sometimes plagued with infestations of inapproriate bacteria because most feedlot operators dose their animals with antibiotics, resulting in unhealthy imbalances of bacteria, as described later in this chapter. A recent solution to the problem is to feed these animals lactobacilli to crowd out less desirable microbes. On the short-term basis on which these animals are kept before slaughter, such probiotic methods may be helpful. However, for long-term healthy maintainance of any herbivore, it would be preferable to maintain a natural assemblege of digestive bacteria.

FIELD MARKS OF GRAM-POSITIVES OF BODY CAVITIES

Most of the gram-positives of body cavities are busy fermenting the food that you eat, especially the more indigestible parts such as cellulose (fiber). They are also enjoying (in moderation) the rich fluids of the nasal passages and vagina. For the most part, they are unobtrusive and produce only subtle field marks. Unfortunately, when the bacteria make themselves too obvious, the situation can become harmful or even pathogenic. For bacteria living so intimately with us and sharing so much in the good things we eat and the good things we are, there is a fine line between coinhabitant and intruder. The fact that you are *not* experiencing anything unusual is, in a sense, the "field mark" of your normal microbiota.

Nasal and Other Respiratory Passages

You produce nasal mucus, then keep it flowing by means of tiny cilia lining the nasal passages; this process prevents bacteria (and viruses and dust) from becoming too firmly established in your respiratory system. In fact the whole respiratory system is a wonderful habitat for bacteria. The lungs are particularly vulnerable because they have a thin lining for the exchange of oxygen into the bloodstream and therefore less of a barrier to exclude bacteria. Some gram-positives of the skin that are found in the nasal area include *Staphylococcus* and *Streptococcus*. *Staphylococcus* and others may help prevent other less desirable bacteria from colonizing. Unfortunately, however, they can turn nasty if opportunities arise, causing staph (*Staphylococcus*) infections, strep (*Streptococcus*) throat, and bacterial pneumonia. Essentially, no news is good news; if you don't notice your respiratory bacteria, then all is going well.

Oral Cavity

Like barnacles that cling to rocks constantly dashed by the waves, bacteria of the mouth are good at sticking to surfaces. The constant swallowing and talking that we do with our mouths makes for a hectic environment indeed. The best strategy seems to be to get in between the teeth or under the gum line and to form cement-like attachment compounds. Yes, dental plaque is a field mark of the gram-positives as well as the *Bacteroides* (chapter 15) of the oral cavity. To some extent, plaque must be considered a natural state, in that all animals have bacteria sticking to their teeth, and it is typically only humans that use toothbrushes, floss, and see the dentist. Our adult teeth are designed to last a lifetime, but average human life spans used to be on the order of 30 to 40 years. Dentistry and tooth care are essential adjuncts to a longer life span.

The fact that you need to brush and floss (to keep your teeth longer than 30 years) is a direct result of the gram-positives of your teeth. The plaque you remove contains 10^{11} bacteria per gram and an adhesive that is also a bacterial product. *Streptococcus, Actinomyces, Lactobacillus, Propionibacterium,* and other gram-positives along with *Bacteroides* form embedded, attached, mixed colonies in plaque. They produce acids (especially lactic acid), which helps to erode tooth enamel, eventually making cavities. They also contribute to the odors of halitosis, or "bad breath." When you brush, you inconvenience your plaque-making bacteria, but those that become dislodged and escape the fate of being washed away quickly (within about 4 hours) reattach and reestablish their mixed colonies. The deepest inhabitants, between gums and teeth, are almost impossible to dislodge by normal hygeine and are among the most odoriferous, producing smelly sulfur compounds. Note that "disclosure tablets" provided by dentists to show how well you are brushing are essentially staining red your plaque-making community of bacteria. Using these tablets is one way to visualize your tenacious oral inhabitants.

Midas Dekkers, in *The Way of All Flesh*, is much more graphic on the topic of tooth decay. He likens the mouth to a "teeming metropolis" full of "slums and alleyways." The bacteria, "like befuddled pub-crawlers at the foot of old cathedrals, . . . urinate against the crowns of your teeth, with a corrosive acid."

Other areas of the mouth such as the cheeks, tongue (especially crevices at the back), and palate are loaded with bacteria, many of them gram-positives. In healthy hosts these tend not to display any obvious field marks other than their contribution to the complex odors of halitosis.

Gastrointestinal Tract

There are relatively few bacteria in the upper regions of the gastrointestinal tract, the esophagus, and the stomach. Food makes a fairly rapid transit through these areas, and the very acidic environment of the stomach makes it relatively inhospitable to bacteria.

The small intestine and large intestine (colon), however, can be literally packed with bacteria. As much as 55% of the weight of the solid content of the large intestine is made up of bacteria, and about 75% of the wet weight of defecated feces is bacterial. In these environments, there is little or no oxygen but abundant food; thus, dozens of species of fermenters flourish. Some are gram-positives, but many are in other bacterial groups (see chapter 2 on methanogenic archaea; chapter 9 on proteobacteria; and chapter 15 on *Bacteroides*). Many of these bacteria digest carbohydrates, including cellulose (fiber), releasing lactic, acetic, and propionic acids as well as carban dioxide as waste products.

In a healthy digestive system, some gas from all these fermenters must be released. What types of gases and in what quantities are addressed in the chapters on methane producers (chapter 2) and the delta proteobacteria (chapter 9). The gram-positives are not major culprits, although they do release carbon dioxide.

The most obvious visual field mark of healthy bacterial activity in the intestines is the forming stool, with 10^{10} bacteria per gram.

Urogenital Tract

In general, the urogenital system of a healthy individual is free of bacteria. The system is regularly flushed out with acidic urine, which is sterile when it is produced. One notable exception is the vagina, which normally contains an abundance of lactobacilli. These acid-producing organisms may help make the environment less hospitable to other bacteria, thus participating in keeping their host healthy. An abundance of *Staphylococcus* in this area can produce toxic shock syndrome. Extra-absorbent tampons, which apparently disrupt the normal microbiota, have been implicated in some cases. Again, no news is good news; the normal microbiota should go relatively unnoticed and produce no obvious field marks.

VIEWING GRAM-POSITIVES OF BODY CAVITIES UNDER A MICROSCOPE

Consider starting with the upper regions of the body. Scrapings, using a toothpick, from the tongue, teeth, and inside of your cheeks ought to produce a slide preparation abundant with tiny rods and coccoids. Another obvious place is a stool sample, of which you need only a small speck, perhaps gathered at the tip of a toothpick. Moisten this with a drop of water to make a slide preparation. Be prepared to find lots of undigested bits of fiber obscuring your view. Nevertheless, the sample ought to have loads of bacteria of all types (some of which are gram-positives). The mucus of the nasal and vaginal cavities makes viewing samples from these areas under the microscope difficult.

CULTURING GRAM-POSITIVES OF BODY CAVITIES

Prepare a plate of moderately nutritious agar as described in appendix A, swab the inside of your mouth or nostril passages, and rub the swab on the agar. Incubate the plate in a warm, dark place for a few days. Expect to find loads of white and yellow colonies, some of which are likely to be gram-positives. An interesting experiment might be to swab your mouth before and after brushing your teeth or using mouthwash. Be sure to swab in exactly the same way both times. The microbes of the intestine are mostly anaerobes and are therefore not easy to culture. Those that might grow on a plate of nutrient agar swabbed with fecal material are not necessarily representative of the community.

CELLULOSE-DIGESTING ANIMALS

Cows, horses, sheep, elephants, hippos, many rodents, some insects (e.g., termites), and most gigantic dinosaurs have something in common: a diet consisting mostly of plant cellulose (fiber). Their food is (or was) indigestible without the symbiotic association of huge, teeming communities of bacteria contained within specially enlarged sections of the digestive system. These bacteria digest the cellulose and produce waste products such as acetate, which are taken up as food by the host. Many of these bacterial symbionts are gram-positives such as *Ruminococcus*, but for a more complete picture of the cellulose-digesting microbes, see also the chapters on methanogens (chapter 2), proteobacteria (chapter 9), and *Bacteroides* (chapter 15).

Cellulose-eating animals may be categorized by the location of their enlarged fermentation chambers. Pregastric types have chambers before or associated with the stomach. These include the ruminants (cows, sheep, deer, and related animals), which are named for the rumen, a large, specialized chamber of the stomach. Much of the barrel-like mass of a ruminant animal is taken up by the rumen, which is packed with bacteria and ciliates (single-celled eukaryotes). The microorganisms themselves may represent almost one-quarter of the animal's body weight. Other pregastric fermenters are listed in table 12.2.

Postgastric animals have chambers after the stomach. The cecal fermenters, for example, have enlarged chambers called ceca between the small and large intestine. These too are packed with microbes and can represent a significant part of the weight of the animal. We humans have a vestige of that organ, the appendix. Included among the cecal fermenters are horses and their relatives, and rodents of all sorts, especially large wood-eaters such as porcupines and beavers. In fact, largeness along with herbivory is such a good indicator of this type of symbiotic digestion that the huge plant-eating sauropod dinosaurs are hypothesized to have been cecal fermenters. Much of the mass of their enormous bodies would have been taken up by digestive organs.

STRATEGIES OF HERBIVORES WITH SYMBIOTIC BACTERIA

The evolution of herbivorous animals has taken interesting turns, many of which are related to the challenges of maintaining bacterial symbionts and getting enough energy from a nutrient-poor diet that is shared with these symbionts. Some of the major

TABLE 12.2. Herbivorous Mammals Containing Symbiotic Digestive Bacteria

Group	Animals with pregastric bacteria*	Animals with postgastric bacteria
Marsupials	Kangaroos	Kangaroos
		Koalas
		Wombats
Primates	Colobine monkeys	Colobine monkeys
	Indriid monkeys	Indriid monkeys
		Howler monkeys
		Lemurs
Edentates	Sloths	
Rodents		Beavers
		Porcupines
		Capybaras
		Naked mole rats
Carnivores		Pandas
Lagomorphs		Rabbits and relatives
Proboscids		Elephants
Sirenia		Manatees
(related to proboscids)		Dugongs
Odd-toed Ungulates		Equines (horses and relatives)
		Rhinoceroses
		Tapirs
Even-toed Ungulates		
Nonruminant	Hippopotamuses	
	Pigs	
Ruminant	Camels	
	Cattle and relatives	
	Sheep	
	Goats	
	Moose	
	Giraffes	
	Deer, antelopes, and relatives	
Cetaceans (related to ruminants)	Whales	Whales

* Animals with pregastric chambers also carry out some degree of postgastric fermentation. Animals that use both chambers extensively appear in both columns.

evolutionary trends are described here, along with examples. Keep in mind that many animals could be listed in more than one category and that the trends are not meant to be mutually exclusive. A project for an amateur naturalist might be to choose an obscure herbivore and to research it in the library (or in the zoo or the wild), keeping this list in mind to see where it fits into this scheme. The strategies of the different herbivores are indirect field marks of their bacterial symbionts.

1. Having an Enlarged Fermentation Chamber

The strategy of having an enlarged chamber for digestive symbionts may seem obvious, given the near impossibility of gaining nutrition from a cellulose-rich diet without bacterial help. However, there are some interesting exceptions—most notably, pandas, which are descendants of omnivorous and carnivorous ancestors but have evolved to consume a diet almost exclusively of bamboo. Pandas have no enlarged chambers for symbionts and seem to be getting along with a bare minimum of helpful bacteria. Pandas appear to pursue a variety of strategies to compensate for the difficulties of digestion. Another exception is the gorilla, which can have a rather leafy diet in the wild but does not have any particular enlargement of the digestive system. However, gorillas are not the exclusive leaf- and stem-eaters that pandas are.

2. Being Big

Many herbivores are large and rotund. They are eating for more than one—in fact they are eating for trillions, when you consider that every mouthful of food feeds both their symbionts and themselves. A fairly reliable hallmark of herbivory in extinct animals is size. Huge, herbivorous megafauna that thrived only temporarily, presumably because of a lack of predators, include giant sloths, giant lemurs, giant beavers, and various giant marsupials. Oftentimes, the arrival of humans to a continent or island marks the demise of a particular giant herbivore in the fossil record. Living megafauna (also called "mega hoof stock" by some zookeepers and defined as being over 1,000 kg) include hippopotamuses, rhinoceroses, elephants, and giraffes (figs. 12.1, 12.2).

Even sharks, notorious flesh-eaters, follow this "rule" of large size as an indicator of a herbivorous diet often accompanied by

FIGURE 12.1. Giraffes (ruminants) and most other megafauna are herbivores that rely on a host of bacterial symbionts for digesting tough plant material. (Photo by Walter Potaznick.)

helpful bacteria. The two largest sharks are the whale shark and the basking shark, both of which pack in a diet of tiny plankton using sieve-like teeth. Much of their bulk (the whale shark is up to 30 ft long) is an enlarged section of the stomach for fermentative food digestion.

Because of having to accommodate digestive symbionts, ruminants cannot be too small. Some of the dwarf antelopes (10 or 11 lbs—the size of hares) may represent the minimum size possible for herbivorous fermentation.

3. Being Slow and/or Aquatic

Many large herbivores with digestive symbionts are also slow, some to the point of lethargy. Sloths are named for this lifestyle, spending most of the day absolutely still and feeding slowly for the remainder. Conservation of limited energy seems to be the outcome of lethargy. Also slow-moving are some (but not all) lemurs, koalas, and pandas. Slowness may have been one of the downfalls of the various extinct megafauna on their first encounters with human hunters.

FIGURE 12.2.   Moose, the largest of the deer, are ruminants that use a high-throughput strategy to compensate for an especially fibrous diet. (Photo by Nancy Marshall.)

However, many slow animals have special strategies to protect against predators. Three-toed sloths are an excellent example: cyanobacteria in their fur give them a camouflaged green tint and perhaps a disguised odor (see chapter 14).

Hippos are semiaquatic, having combined resting with the benefits of being buoyed up by water. Some of the herbivorous ancestors of whales and manatees may have taken that strategy to an extreme, resulting in a completely aquatic life in their descendants. Whales are of a ruminant (cow-like) lineage, and manatees are descended from elephant-like ancestors (plate 52).

4. Being "High-Throughput"
An alternative to a large, vat-like physique is to have a fast-moving pipeline that can process large quantities of food in such volume that both host and symbionts can be supported. This seems to be the strategy for many equines (horses and their relatives), which are fairly streamlined and fast moving in spite of a cellulose diet. Digestion moves along more quickly than in larger herbivores; the

end result can be seen in the feces, which often have recognizable plant materials that are only partly digested.

Pandas are the most striking example of high-throughput herbivory. An individual may eat 30 to 40 pounds of bamboo per day (a challenge for zookeepers) and deposit 15 to 20 pounds of feces (yet another challenge) loaded with large leaf and stem parts. They are considered to be among the most inefficient of the herbivores and yet are able to maintain their large, slow-moving bulk.

5. Eating Twice

Many herbivores have some way of reingesting food from their fermentation chambers. The phenomenon can be difficult to observe, however, even in captive animals, and may be much more widespread than has been documented. In ruminants, eating twice is called rumination or cud chewing. Cattle, moose, giraffes, deer, and others do this by regurgitating wads of partly digested, symbiont-rich plant material after it has been fermented in their stomachs, and then rechewing to gain additional nutrients. The long necks of giraffes sometimes reveal the progress of such boluses of food from the stomach up to the mouth via muscular contractions.

Some (and perhaps many more than reported) postgastric fermenters perform coprophagy, the consumption of their own or others' feces. All lagomorphs (rabbits) and rodents do it; however they can be quite secretive about the activity. In both these groups, feces are of two types: (1) primary feces, which are soft, moist pellets full of symbionts; and (2) secondary feces, which are the hard, dry pellets familiar to those who have cleaned rabbit or rodent cages. Primary feces are consumed as they are being produced, often with a quick motion that looks like grooming and is difficult to observe. It has been reported for beavers, porcupines, and capybara, and it probably occurs in all wild herbivorous rodents. It is also reported in some lemurs and in manatees. A fairly definitive experiment that has been used for captive animals is to attach an "Elizabethan collar" to prevent coprophagy. Under these circumstances, it is possible to observe uneaten primary feces to confirm the behavior. Some captive rodents with Elizabethan collars experience nutritional deficiencies as a result of not being allowed to eat their primary feces.

6. Being Social

It has been a long-standing hypothesis that the sociality of termites evolved around the pressures of maintaining digestive symbionts and subsisting on an extremely nutrient-poor diet. Offspring are born sterile and receive their inoculum of symbionts and digested wood from the primary feces of attending workers. Reproduction is rationed to only one reproductive pair because it requires an excess of protein, one of the most limited nutrients in a wood diet. The queen termite receives extra protein from the primary feces of workers. Thus, a social organization maintains the symbionts and helps allocate limited protein to reproduction.

Can the termite strategy be found in any herbivorous mammals? Perhaps, especially in species in which one breeding pair is tended by several generations of nonreproductive offspring. Naked mole rats seem to be the best mammalian example (fig. 12.3). Their cellulose-rich diet of plant roots may have contributed to the evolution of their termite-like social structure. Certainly the eating and feeding of symbiont-rich feces plays an important part in their colonial behavior.

FIGURE 12.3. Naked mole rats have digestive symbionts in their enlarged ceca.

To a lesser degree, extended, cooperative families of several generations may be seen in other rodents, such as beavers. Close observations of other herbivores may reveal similar patterns. A parental pair tending to more than one generation of offspring, with older siblings assisting younger ones and with sharing of primary feces, may be an indication of herbivorous sociality. Beavers, and even herds of ruminants such as cows, have aspects of this family structure. Hippos, too, may be good candidates, because they live in groups with overlapping generations and have some interesting behaviors based on coprophagy.

7. Being Solitary

In contrast, many large, slow-moving herbivores such as sloths, porcupines, and koalas are solitary, perhaps a strategy to restrict competition for limited nutrients. Weaning of offspring is often accompanied by an inoculation of symbionts via primary feces. Koala mothers, after weaning and before departing for solitary lives, defecate huge amounts of soft green feces which are devoured by the young. Indeed, the pouch of the koala opens toward the anus (unlike that of other marsupials) so that the offspring can get better access to the bacteria-rich defecation.

8. Scavenging Protein

One of the most limited and necessary nutrients in a cellulose-rich diet is protein. Protein is essential for reproduction, and a lack of it can increase gestation length, decrease the number of offspring, and delay the onset of maturity. Fortunately, bacterial symbionts can be quite efficient in providing useable nitrogen and protein. They do it in three majors ways: (1) by fixing atmospheric (gaseous) nitrogen into forms that can be incorporated by the host; (2) by recycling nitrogenous wastes such as urea; and (3) by serving as protein-rich food themselves, succumbing to digestion by the host.

Furthermore, many herbivores seek out protein by eating insects or feces, by scavenging dead animals, and, in captivity, by being "open-minded" about the more protein-rich diets provided by humans. Thus, many captive herbivores can survive well on zoo diets that are much richer in protein than a typical wild diet. Sometimes the advantages of this type of diet are at the expense of the full diversity of symbionts and the full range of behaviors concerning those symbionts.

9. Being Able to Eat Toxic Plants

Symbiotic bacteria can also detoxify some of the plant compounds that usually serve as antiherbivore defenses. This is why, for example, koalas can thrive on poisonous eucalyptus leaves with almost no competition from other herbivores. Even so, some biologists have surmised that the lethargy of koalas may be a druglike effect of the toxins. Some species of eucalyptus contain cineol, which lowers blood pressure and body temperture, relaxes muscles, and (in high doses) can even stop respiration. Koala brains seem to develop to a smaller size than those of comparable animals, presumably due to the debilitating effects of the toxins. Koalas have no ectoparasites (such as ticks and fleas) because their skin is so saturated in toxic eucalyptus compounds.

10. Being Selective versus Unselective

Some ruminants, such as domestic cattle, are fairly unselective in their choice of plants. This behavior enables them to waste little energy in searching, "trusting" that all plants will be fermented somehow in their ruminant stomach chambers. On the other hand, some herbivores are quite selective in their diet, using energy to find the most nutritious plant parts. Seeds and fruits can supplement and improve a leafy diet, as can insects or scavenged animals.

Some animals, such as the toothed whales, seem to have evolved from herbivory through omnivory into carnivory. All whales (cetaceans) seem to be related to ruminant herbivores, and they retain the complex stomachs of that lineage. In the case of baleen whales, microbes of both the ruminant stomach and the enlarged cecum aid in the digestion of planktonic organisms. Microbes in the stomach of smaller toothed whales also aid in digestion, although the diet is not particularly herbivorous. Larger predatory toothed whales, however, seem to have lost the necessity for digestive microbes.

SYMBIONTS AND DOMESTIC HERBIVORES

Ruminant animals such as cows, sheep, and goats are of such great economic importance that they have been subjected to extensive experimentation to determine optimal diets for the best output of milk and meat—strategies that are not always compatible with what is best for the bacteria. Some experimentation on ruminants has been done on fistulated animals, especially cows. These animals have had a hole made in their sides so that rumen fluid and symbionts can be removed and analyzed.

For maximum milk production, cows are fed a diet high in grain (rich digestible carbohydrates and protein). However, this diet can increase lactate production in the stomach and may cause an excess growth of lactate-using bacteria. This change can result in digestive conditions being too acidic; sodium bicarbonate is sometimes added to high-grain diets as a deacidifier. Such diets can also encourage the growth of potential pathogens such as *E. coli*, as well as causing other digestive problems. Even for vigilant farmers, such conditions may occur in cattle turned loose in a pasture overly full of delicious, protein-rich legumes. In contrast, a high-fiber diet of grass or hay, which perhaps better approximates the diet of wild bovines (and their symbionts), does not yield large quantities of milk and is therefore not practical as the sole diet for dairy animals. A careful balance is necessary for maintaining healthy milk-producing cows.

Remarkably (at least to this microbiologist), animals to be used for meat or wool are sometimes treated with antibiotics to deliberately remove some bacteria along with the larger eukaryotic digestive microbes, the ciliates, which consume the bacteria and bacterial products that would otherwise be channeled into making more animal tissue. Supposedly this is done without harm to the animal, although a lack of ciliates is far from a normal condition in a ruminant.

Feedlot animals are sometimes dosed with antibiotics during their final days for a variety of reasons other than removing competing symbionts; one such reason is to provide prophylaxis against possible diseases that might spread rapidly in crowded conditions. Use of antibiotics sometimes leads to an overgrowth of bacteria such as *E. coli*, which is usually a minor member of the microbial community. This is one of the ways in which meat can become contaminated, resulting in food poisoning.

Ruminants that completely lack digestive symbionts, as a result of treatment with a large does of antibiotics, are in trouble. If the animals are not about to be slaughtered, farmers sometimes use rumen fluid from healthy animals to reinoculate the system. Also, calves that are raised in isolation on sterile feed are not properly inoculated with symbionts. The natural condition is for weaned animals to consume symbionts in cud (regurgitated material from the parents), which is deposited on grass during feeding. Some calves are isolated deliberately to produce mild-flavored veal. Abandoned calves must receive a dose of rumen fluid to survive past calfhood.

One of the ironies of ruminant-style digestion is that it is more efficient with a fiber-rich, nutrient-poor cellulose diet than with one that has been supplemented with grains or other protein- and carbohydrate-rich foods. The microbes in a ruminant's stomach essentially compete with the animal for the best nutrients and win much of the time, having more direct access to the food. The richer the food, then, the happier and healthier certain bacteria and protists are. Fibrous (relatively nutrient-poor) foods digested twice, slowly, in typical ruminant fashion are more likely to result in nutrients that can be absorbed more equitably by both bacteria and host.

Horses are an interesting contrast to cattle. They are not ruminants but rather hindgut fermenters that digest cellulose in their ceca and intestines. They need higher-quality diets with more protein and easily digested carbohydrates than ruminants do. Intestinal fermentation might be considered to be a less efficient method of using cellulose. Horse intestines are full of helpful bacteria and ciliates, but their microbes have briefer access to nutrients because food travels more quickly through horses and out. However, horses seem to compensate for this. Horses that feed on their own in a pasture tend to be rather selective grazers for more nutritious plants, and their "high-throughput" method of digestion is indeed quite efficient. Although horses are rarely used for meat or milk, their diets enable some of them to gallop at high speed while carrying a rider, a phenomenon rarely seen in cattle.

Many wild ruminants, including giraffes and moose, are (like horses) selective eaters, preferring richer foods. They use their intestines more like horses and other equines do and have relatively small rumen stomachs for their size.

One element of domestic cattle's diet, silage is already partially digested before they eat it. Silage consists of chopped corn, grass, clover, and other plants compressed and fermented by lactic acid–producing bacteria within a silo; the process is similar to that of making pickles (chapter 10). The resulting product is a succulent winter feed for cattle. The process of producing silage may be thought of as part of a cellulose-digesting continuum. In this case, the problem of breaking down cellulose is solved in part by processes taking place outside the cow, although with similar bacteria to those found within the animal. The challenges of cellulose as a food source are such that the same basic solution appears over and over—using fermentative bacteria (or in some cases fungi) to get edible, digestible material.

HERBIVOROUS FISH

Ichthyology is considered to be an extremely complex branch of zoology—even by many zoologists themselves. Perhaps it is because we humans have so little personal experience on which to base hypotheses and observations on a completely watery existence. Therefore, although table 12.3 lists species with symbionts, for the benefit of amateur naturalists who snorkle or keep aquaria, very few generalizations about fish are attempted here. The fish listed in table 12.3 are plant and algae eaters. They have enlarged intestines containing digestive symbionts, a relationship that has not been a major focus of ichthyology. Much work remains to be done on fish herbivory. Sufficient quantities of plant or algal food are difficult to obtain in many fish habitats; the best place for herbivores seems to be in the tropics, and the most accessible of these locations are coral reefs.

The surgeonfish *Acanthurus*, one of the few well-studied fish with symbionts (plate 53), is especially noteworthy for having the largest bacteria known, the gram-positive *Epulopiscium* (read more about this giant, visible as a large speck to the naked eye, in the introduction). Possibly the stability and constant flow of nutrients through the intestine were the selection conditions for the evolution of such a large cell. In general, the digestive systems of herbivores across the animal kingdom contain some of the largest examples of both prokaryotic and eukaryotic microbes.

TABLE 12.3. Fish Containing Intestinal Algae and Plant-Digesting Symbionts

Tropical freshwater fish
  Cyprinidae
    ruby shark
    red-finned shark
  Citharinidae
    Distichodus
  Loricariidae (many of the catfish)
    Otocinclus
    Farlowella
    Hypostomus
    Ancistrus
  Gyrinochelidae
    Gyrinocheilus (algae eater)
  Cichlidae (some)
    Labeotropheus
    Melanochromis
  Osphronemidae
    Osphronemus (gourami)
  Loricariidae
    Ancistrus (bristle nose)
  Anostomidae
    Abramite (marbled head stander)
  Serrasalmidae
    Colossoma (pacu)
    Metynnis
    Mylossoma (silver dollar)
  Characidae
    Poptella (silver dollar tetra)

Tropical marine fish (reef fish)
  Pomacanthidae (some)
    Pomacanthus (e.g., blue ring angelfish)
  Acanthuridae (all)
    Acanthurus (tangs, surgeonfish)
  Siganidae
    Lo (fox face)
  Scaridae (some) (parrot fish)
  Blenniidae (some)
  Kyphosidae (some)

Herbivorous fish can sometimes be difficult to keep in aquaria, especially if they do not tolerate the dried foods provided to more omnivorous fish. Letting algae or plants grow in the tank is not the best solution, as it can disrupt the ecological balance. For example, *Acanthurus* develops nutritional deficiencies if it is not supplied with occasional plants or algae and if unconsumed items are not removed immediately.

HERBIVOROUS BIRDS AND REPTILES

Exclusively leaf-eating birds are extremely rare. The best example is the bizarre tropical bird the hoatzin, which has an enlarged crop full of bacteria. It is the only pregastric fermenter of the birds. It has been described as smelling strongly of fermentation (actually a sort of cow manure odor) and has an enlarged sternum to support its crop. Other birds that consume a significant amount of leafy material include grouse, ptarmigans, ostriches, and rheas—all of which are postgastric fermenters and have the characteristic rotund shape of many herbivores. Canada geese that graze on lawns are postgastric fermenters; they use the high-throughput strategy described earlier for pandas, which explains why their feces are so green.

Among the herbivorous reptiles with postgastric symbionts are iguanas, green sea turtles, some land tortoises, and most likely the gigantic sauropod dinosaurs, long extinct but deserving of mention here. Can we really draw conclusions about the details of an animal's diet from the fossil record? Dentition studies indicate that sauropods were herbivorous, and the enormous vat-like bodies are just right for holding a colon full of digestive bacteria. Implications for sauropod behavior are intriguing, since maternal care or at least proximity is required for passing along digestive symbionts to offspring.

HERBIVOROUS INVERTEBRATES

It is an easy (almost foolproof) rule of thumb that if an animal munches on wood, it has symbionts that help it digest the cellulose. Most famous of the wood munchers are termites, but almost any rotting tree is also loaded with scurrying beetles, millipedes, and isopods. Many of these invertebrates do indeed have symbionts, but a corollary to the rule of thumb is that the sounder the wood, the more likely it is that the symbionts are internal, bacterial, and of primary importance. If fungi are living in or on the wood, they are in effect predigesting it, precluding a need for extensive bacterial symbionts. Some tropical termites even "garden" fungi to use the by-products of fungal digestion of plants.

Invertebrates in marine waters display some interesting variations and exceptions. Certainly any invertebrate that is "fouling" wood is a good candidate for having digestive symbionts. Indeed, *Teredo*, the shipworm (actually a mollusc), has digestive symbionts that are proteobacteria (see chapter 8). Algae-eating echinoderms also have symbionts, including gram-positives, as do some herbivorous crustaceans. There are some bacteria-free herbivorous isopods and amphipods (including two terrestrial species). These crustaceans have enzymes that can partially digest cellulose, making them quite exceptional in the animal world.

Cellulose Digestion in Humans

Humans cannot digest cellulose (fiber) very efficiently, but even vegetarians do not rely solely on diets consisting of the tough leaves and stems of plants. We tend to be omnivores—consuming roots, nuts, fruits, seeds, and in some cases other animals or animal products. Thus, there is not such a crucial need for large symbiotic communities of cellulose digesters. If there were, we might by necessity be naturally more rotund, more sedentary, and perhaps 10 to 20% heavier due to the weight of microbes. Humans have a remnant of an ancestral cecum, the tiny appendix, located between the small and large intestines. Perhaps it was more active in our distant ancestors.

Cellulose-digesting bacteria, many of them gram-positives, thrive on the fiber in our large intestines. About half of the fiber people in developed countries consume is digested by microbes. The resulting microbial waste products do not, however, seem to be major or essential components of our omnivorous diets, comprising only about 2% of the nutrients in "Western" diets. There has been some speculation about whether indigenous "non-Western" peoples with highly plant-rich diets might have populations of bacteria that are better able to digest cellulose, use urea as a source of nitrogen, and fix nitrogen. For example, some highlanders of Papua New Guinea have very high-fiber, protein-poor diets and yet do not appear to have nutrient deficiencies. It may be that a somewhat different assemblage of bacteria is selected by different diets. A sudden change in diet, such as a feast of meat can cause enteritis necroticans (an infection of the intestines) in New Guineans as well as other people with vegetable diets, and may have afflicted medieval Europeans. The abrupt change can allow a toxic increase of gram-positive *Clostridium*.

HERBIVOROUS AND CARNIVOROUS PLANTS

The nutritional strategies of some plants are a variation on the use of digestive symbionts by animals. These include some of the tropical tank bromeliads and insectivorous plants of bogs.

Tank bromeliads (see also chapter 14) provide habitats for cyanobacteria and a whole community of other microscopic and macroscopic organisms. Some tank bromeliads may benefit from a sort of passive fermentation, described as herbivory combined with carnivory by researcher J. H. Frank. Both *Cartopsis berteroniana* and *Brocchinia reducta* grow in nutrient-poor conditions, the former as a rootless epiphyte, the latter in poor soil. Plant materials and insects fall or are blown into the tanks (water-filled enclosures at the base of leaves; plate 54) and are decomposed by the community of bacteria (including gram-positives) and fungi. This decomposed material apparently comprises a significant part of the nutrition of these plants.

Insectivorous plants of bogs supplement their nutrients with nitrogen-rich protein digested from insects. These insects are trapped by various sticky substances or triggering mechanisms, or they simply drown in the water-filled pitchers (plate 55). Bogs and other acidic, waterlogged environments are usually poor in nitrogen compounds. Both *Sarracenia* and *Darlingtonia* (pitcher plants of the East Coast and West Coast of the United States, respectively) have bacteria including fermenting gram-positives in their water-filled reservoirs by which they digest stranded insects. *Sarracenia* supplements this process with enzymes, while *Darlingtonia* apparently digests with bacteria exclusively.

Bladderworts, genus *Utricularia*, are native to bogs or nutrient-poor waters. These carnivorous plants trap small organisms in bladder-like pouches of their leaves. Bacteria in the bladders assist in digesting some of the trapped organisms (e.g., insects), although certain organisms, such as rotifers, apparently stay alive and take up residence.

VIEWING GRAM-POSITIVE SYMBIONTS UNDER A MICROSCOPE

Depending on where you live, try digging up some termites, such as *Reticulitermes* on the East Coast or *Zootermopsis* on the West Coast, or one of several species in the southern latitudes. (More detailed instructions on working with termites are given in chapter 16.) Then follow the directions in appendix C on microscopy to see an amazing array of microorganisms, including large protists (flagellates).

If you live near a large agricultural research facility or land grant university, you may be able to view a fistulated cow. These animals have a hole cut into their side to expose the rumen contents and allow researchers to remove small volumes to view under microscopes. They may allow you to have some too.

If you are able to obtain a sample, be aware that rumen microorganisms are sensitive to oxygen. Fill a jar to the top with rumen fluid, seal it well, and rush it home for immedrate viewing. You could also visit a slaughterhouse and request a jar of rumen contents from a freshly killed animal. As with termites, expect to see lots of protists (ciliates) in addition to gram-positives and other bacteria, such as methanogens (chapter 2).

You may also want to view samples from the water-filled tanks of bromeliads or from the reservoirs of carnivorous plants. Try plucking out a decaying insect from an insectivorous plant to view the bacteria associated with that decay.

CULTURING GRAM-POSITIVE SYMBIONTS

The symbiotic cellulose digesters tend to be very sensitive to oxygen, and it is therefore not a simple home project to culture them.

SUMMARY: FIELD MARKS AND HABITATS OF GRAM-POSITIVE SYMBIONTS OF ANIMALS

Humans

- skin
  - active sebaceous glands and some adult body odor (*Propionibacterium*; high GC*)
  - smelly feet (*Brevibacterium*; high GC)
  - other distinctive body odors (*Staphylococcus*; low GC)
- dental plaque (*Actinomyces, Propionibacterium* [high GC]; *Streptococcus, Lactobacillus* [low GC]) (note that *Bacteroides* are also present in large numbers; see chap. 15)
- nasal and respiratory passages (*Staphylococcus* and *Streptococcus*; low GC)
- gastrointestinal tract (many different gram-positives, along with methanogens [chap. 2], proteobacteria [chap. 9], *E. coli* and other coliforms [chap. 8], and *Bacteroides* [chap. 15])
- urogenital tract (*Lactobacillus*; low GC)

Cellulose-digesting animals (e.g., cows, horses, sheep)

- gastrointestinal systems (many different gram-positives, along with methanogens [chap. 2], proteobacteria [chap. 6], and *Bacteroides* [chap. 15]) (see table 12.2)

Carnivorous and herbivorous plants

- "tanks" of bromeliads and reservoirs of pitcher plants (a variety of gram-positives and other organisms)

* High concentrations of the nucleic acids guanine and cytosine.

Introduction to the Cyanobacteria

If bacteriophilic humans were able to explore ancient Earth for field marks of bacteria their field guide could be much simpler than this one, because it would not need to explain the difference between bacterial and eukaryotic field marks. Lots of bright, greenish-blue slime growing in moist sediments, floating on water, or clinging to rocks would have been infallible indicators of cyanobacteria (the dominant photosynthesizers). It was a bacterial world.

Now, however, eukaryotic photosynthesis prevails. Since the evolution of eukaryotic algae about 2 billion years ago, an enormous diversity of photosynthesizing organisms has evolved. On modern Earth, plants occupy nearly every patch of moist, rich soil, and bright-green slime is likely to be some type of eukaryotic algae. The trick to finding blooms of cyanobacteria (also called blue-green bacteria) is to look in places that are not so pleasant for eukaryotes. From a human point of view, these are sometimes considered extreme environments because they deviate in some way from those conditions we consider to be ideal. Good places to search include hot springs, acid or alkaline lakes, hypersaline waters, dry desert crusts, and even bare Antarctic rocks.

A New Type of Photosynthesis

The great evolutionary innovation of cyanobacteria was oxygen-generating photosynthesis. It probably evolved about $3^1/_2$ billion years ago and, once established, became the dominant metabolism for producing fixed carbon (in the form of sugars) from carbon dioxide. All types of photosynthesis use light energy to fix carbon, but unlike earlier photosynthesizers, cyanobacteria used water as a source of hydrogen and electrons for converting carbon dioxide into sugar:

$$\text{light energy} + CO_2 + H_2O \rightarrow \text{sugars (food)} + O_2 \text{ (waste product)}$$
$$(\text{e.g., } C_6H_{12}O_6)$$

The waste product is oxygen, which, although it is toxic, is a gas and therefore can diffuse away from the cell. Oxygen is toxic because it oxidizes molecules, including those that organisms are made of. When oxygen began to accumulate due to cyanobacterial photosynthesis, other organisms either moved away or evolved mechanisms for coping with oxygen; respiration using oxygen was one of these mechanisms.

Before the evolution of cyanobacteria, photosynthetic organisms had used hydrogen sulfide (H_2S) as a source of hydrogen and electrons, producing sulfur or sulfate as a waste product, or they had used organic acids such as acetate as hydrogen and electron donors, producing smaller organic acids as wastes. Not only are those wastes (sulfur compounds and organic acids) somewhat more difficult to get rid of than oxygen, but the starting compounds (hydrogen sulfide and organic acids) are more difficult to acquire than water, which is nearly everywhere. In short, water-using, oxygen-producing photosynthesis was such a successful variation that it quickly became the dominant mechanism for fixing carbon (making sugar). It has been, and continues to be, the source of most productivity on Earth, as evidenced by the enormous biomass of plants, algae, and cyanobacteria—all of which use oxygen-generating photosynthesis. Indeed, plants and algae themselves evolved from cells that acquired cyanobacteria as symbionts about $2^1/_2$ billion years ago. Thus, even a tree is a manifestation of its symbiotic cyanobacteria—or former symbionts—which are now integral parts of the plant cell called chloroplasts.

FIELD MARKS OF CYANOBACTERIA

Oxygen is a field mark of cyanobacteria and, for that matter, all other oxygen-generating photosynthesizers, such as algae and plants. If you are observing a mat of cyanobacteria on a sunny day, you may see oxygen bubbling at the surface. Oxidized waters, sediments, and atmospheric oxygen are also hallmarks of oxygen-generating photosynthesis. With the exception of a few deep muds, stagnant waters, deep springs, and animal digestive tracts, the Earth is an oxidized world. True anaerobes are now restricted to those dark environments, having predominated 3 to 4 billion years ago, before oxygen began to accumulate. It may be easier to identify anaerobic habitats because they seem so strange to us: smelly, dark, often enclosed. We take the oxygen-rich habitat for granted. However, oxygen is one of the distinguishing characteristics of our planet, along with its distance from the sun and the presence of liquid water. Oxygen and all the oxidized sediments and oxidized waters are, in a sense, planet-wide field marks of cyanobacteria.

Ancient limestones are also a hallmark of cyanobacteria. Most carbon on Earth is not contained within the biomass of organisms but is deeply buried as limestone that originated as calcium carbonate, precipitated by the activities of photosynthesizers. The next time you see a building constructed from limestone, look at it closely. Limestone that is more recently deposited (<500 million years ago) may be full of marine animal fossils and is probably a product of eukaryotic photosynthesis. But ancient limestone, more than 2 billion years old, is a product primarily of cyanobacteria.

CHAPTER 13

Cyanobacteria of Aquatic and Terrestrial Habitats

Wherever there is moisture or the potential for sporadic moisture in an area reached by light, there is the possibility of finding cyanobacteria. This chapter discusses some of the best habitats for seeing unobscured cyanobacteria. As is true for most bacterial groups, these tend to be "extreme" from the viewpoint of eukaryotes. Although cyanobacteria are ubiquitous worldwide, environments that lack eukaryotes are the most rewarding. Thus, this chapter focuses on dimly lit waters, desiccating splash-zones, sulfury mud flats, dry desert-like soils, bare rocks, and even glaciers. Cyanobacteria have also left striking fossils (such as stromatolites) from the glory years before eukaryotes became so abundant; instructions for viewing and interpreting some of these fossils are included as well.

FIELD MARKS AND HABITATS OF FRESHWATER CYANOBACTERIA

Cyanobacteria of fresh waters are often in competition with eukaryotic algae of all sorts, but especially diatoms and green algae (variously named pond scum, pond silk, frog spittle, blanket algae, and water bloom). There are some general rules for making reasonably confident identifications of cyanobacteria by eye. Often, however, communities are so mixed that even if you've located some cyanobacteria, there may be eukaryotes as well. The trick is to seek out marginal environments, the more inhospitable to eukaryotes the better. Favorable conditions for identifying cyanobacteria include low light, extremes of pH (very acidic or very alkaline), periodic drying, and extremes of temperature. Also, since many cyanobacteria can fix nitrogen (essentially making their own fertilizer), they can live in nitrate-poor waters.

Another bit of advice is not to expect cyanobacteria to be truly blue-green or green, even though green chlorophyll is one of their essential pigments. Grassy-green colors are hallmarks of eukaryotes, whereas cyanobacteria are often black or blue-black. Sometimes they display their secondary pigments: yellow and brown carotenoids or red and purple phycoerythrins. In fact, cyanobacteria can be almost any color except the green that you might expect in a photosynthesizer. The following pages give more specific descriptions of what you might observe in various types of fresh waters. In some cases possible genus names are suggested, but confirmation with a microscope is needed to be completely confident about an identification. Many of these genera are also found in marine waters, terrestrial habitats, and as symbionts (see chapter 14).

PONDS, LAKES, STREAMS, AND RIVERS

Bright green (grass-green) silky or hairy-looking pond scum that is floating, submerged, or attached to rocks or logs is likely to be eukaryotic green algae (plate 56). If the water is rich in nutrients due to run-off of fertilizer from lawns or gardens, then green algae are probably more prevalent than cyanobacteria.

Look for cyanobacteria in the shadows of stones, bridges, or other objects. Look for them just above the water line or in a splash zone, where they might be dried out due to evaporation. If they are dry, they may look brown-black and crusty. On re-wetting, they may become somewhat less black and more blue-green, though browns, reds, and oranges may be their predominant colors, in which case it may be more challenging to identify your sample as cyanobacteria. Or look for cyanobacteria deeper in the water, where the light is dimmer. Depending on how clear the water is, they may be found either attached to a substrate or floating in diffuse masses centimeters to meters below the surface of the water. Deeper cyanobacteria are often reddish-purple, whereas those closer to the surface are more likely to be yellow-brown (plate 57).

Although almost any cyanobacteria can become bleached out,

revealing yellow-orange-red pigments, some species have more of a tendency to reveal these colors. A reddish tint to the water may mean that *Oscillatoria* is present (although many other things can make water reddish, including oxidized iron). *Gloeocapsa* can make reddish-orange floating gelatinous masses.

If cyanobacteria bloom in great numbers, they may then decay with a detectable odor, which is sometimes described as fishy. Masses of green algae, however, can also produce off odors; in general, a decaying anaerobic environment exudes a complexity of smells.

Filamentous Cyanobacteria

Filamentous cyanobacteria of fresh waters include species that form globular, mucous colonies that float or are attached to vegetation or other substrates. Individual colonies range from microscopic to 10 centimeters in diameter (plate 58). Common genera include *Nostoc*, which makes an especially firm, rubbery, blue-green-brown glob, and *Anabaena* and *Gloeotrichia*, which form somewhat looser structures. The name *Nostoc*, coined by the Renaissance physician Paracelsus, refers to the fact that colonies can look like nasal discharge. Such colonies also have many colloquial names, including "philosopher's stone," "mare's eggs," "witch's (or fairy's) butter," "fallen stars," and "star spittle." Some French and German names include "crachat de lune," "purgations des étoiles," and "sternschnuppe."

Films or fuzz, often yellow-brown on submerged vegetation, may be *Schizothrix* or *Stigonema* (see plate 57). Masses of filaments floating on the water or submerged might be *Microcystis* or *Plectonema*. Some filaments form little floating clumps: *Aphanizomenon* can look like bits of chopped grass floating in the water, and *Gloeotrichia* has been described as looking like buff-colored tapioca in the water.

Other filamentous cyanobacteria form dark mats or felts that are often quite slippery when exposed on wet rocks or damp soils on the shoreline. If you gently tear apart a mat, you can see that it is made of tiny lint-like fibers. These felt-makers include *Phormidium*, *Oscillatoria*, *Spirulina*, *Lyngbya*, *Stigonema*, *Calothrix*, and *Microcoleus*. In lakes or marshes that are rich in calcium carbonate, cyanobacterial mats may become encrusted with this mineral.

Coccoid Cyanobacteria

Coccoid (little round) cyanobacteria usually have less visible field marks than do filamentous species unless they are gathered together in colonial form. *Gloeocapsa* is sometimes visible as a reddish-orange colony. *Merismopedia* can form thin sheets of embedded cells, although this is a somewhat rare field mark.

HOT AND MINERAL SPRINGS

Eukaryotic algae are not as tolerant of thermal waters as are cyanobacteria. At temperatures greater than 60°C (140°F), no eukaryotes will be present; temperatures up to 75°C (167°F) cyanobacteria can tolerate. Therefore, blackish-blue, green, or some yellow colors in and around sufficiently hot springs almost certainly indicate cyanobacteria. (For a more complete interpretation of the colors and temperatures of hot springs, see chapters 1, 3, and 17.) If calcium carbonate (travertine) is being deposited, look for cyanobacteria coloring that mineral. Alkaline springs have a greater abundance and diversity of cyanobacteria than acidic ones. In sulfur-rich springs, expect to see variations of blackish blue-green cyanobacteria alongside purple sulfur bacteria, perhaps white *Beggiatoa* bacteria, and other sulfur, thermal-tolerant bacteria (see chapter 9, plate 40). Keep in mind that eukaryotes will occupy the more brightly lit parts of the substrate; look for cyanobacteria in the shadows and crevices.

WASTEWATERS

Waters polluted with metals such as manganese, copper, cobalt, and zinc can be cleaned by running them through or across microbial mats. Both cyanobacteria and eukaryotic algae have a natural tendency to bind metals to their sheaths and walls. Indeed, most types of bacteria do this, but it is often easier to get large masses of photosynthesizers for commercial applications. Some cyanobacteria also degrade hydrocarbon pollutants. Quilts of cyanobacteria and algae are sometimes grown on a mesh material and used to filter contaminated water in wastewater plants. Quilts full of metal may then be collected and discarded, leaving behind relatively clean water.

In some mining operations, cyanobacteria and algae are used to capture metals from wastewaters, thus clearing the waters and scavenging metal. The technique is used most extensively in copper mining but also has been applied in uranium and gold mining. It is not easy to get tours of mining operations, especially those that involve strategic metals; however, if you have an opportunity to visit a copper mine, you may see cyanobacteria and algae in use to treat wastewater.

FIELD MARKS AND HABITATS
OF MARINE CYANOBACTERIA

As in fresh waters, cyanobacterial field marks in marine waters may be obscured by a prolific growth of algae (green, red, or brown seaweed). Tropical and nutrient-rich waters are the least likely to show cyanobacterial field marks. Exceptions are hypersaline waters, which cyanobacteria tolerate well. In general, look for marine cyanobacteria according to these guidelines:

1. There are many planktonic cyanobacteria of open waters, but they are often not visible. When blooms do occur, browns and reds are usually the predominant colors, resulting in an effect that may have given the Red Sea its name.
2. Low-light conditions (deeper water) are somewhat less likely to support large seaweeds and more likely to support cyanobacteria.
3. Hypersaline conditions due to the evaporation of seawater may support more cyanobacteria than algae.
4. Porous rocks such as sandstones and carbonates are more likely substrates for cyanobacteria than smooth, harder rocks. However, smooth rocks may support marine lichens, some of which are cyanolichens (see chapter 12).
5. Areas such as sandbars that are left completely dry at low tide are more likely to have cyanobacteria than are moister areas.

PLANKTON
Cyanobacteria are an important part of marine plankton and can be found by microscopic observations of material gathered in plankton nets. However, they do not usually bloom to the point of being

detectable by the naked eye. Farooq Azam, a marine biologist, has shown that many planktonic organisms form clusters, clinging to particles of detritus and cycling nutrients to each other. Each miniature community might include a diatom, several other protists, 1,000 heterotrophic bacteria, 100 viruses, and about 100 cyanobacteria— all clinging to each other and to a particle of debris. If the particle is visible (even as a speck), it is called "marine snow." (For additional information on some of the heterotrophic bacteria of marine snow, see chapter 8). Phenomena such as this miniature community make it especially difficult to count oceanic bacteria and to account for their activities. It has been suggested that bacterial participation in the carbon cycle of the ocean may be underestimated by as much as 99%.

A planktonic species of *Oscillatoria* (formerly called *Trichodesmium*) is more easily observed. Several long filaments of this cyanobacterium clump together in reddish bundles that float on the water, looking like bits of chopped up grass—also referred to as "sea sawdust."

HYPERSALINE ENVIRONMENTS

If a shallow marine environment is subject to extremes such as hypersalinity and desiccation with or without a periodic inundation with salt water or fresh water, then most eukaryotes will be scarce and cyanobacteria may sometimes be found in blooms. For example, lagoons where salt water evaporates, leaving behind salt-saturated water and salt crystals, may contain genera of filamentous cyanobacteria such as *Microcoleus*, *Lyngbya*, and *Spirulina*. Note that the eukaryotic alga *Dunaliella* also thrives in hypersaline waters and even on salt crystals. Halophilic archaea may be there as well (see chapter 4). The cyanobacteria are likely to be a deep blue-green-black (see plate 15). Both the halophilic archaea and *Dunaliella* are likely to be pink or red. Commercial salt works where salt is being evaporated in pools can be wonderful places to see blooms of salt-loving cyanobacteria. Some inland seas (for example, Great Salt Lake in Utah) can be places to look for hypersalinity and cyanobacteria. Indeed, any highly mineralized lake or pond, whether alkaline or acidic, is likely to be inhospitable to most eukaryotic algae and appealing to some cyanobacteria.

ROCKY SHORES

Shells are often colonized by both cyanobacteria and algae as well as by many animals. Some of the cyanobacteria produce acids that dissolve the shell and create crevices in which the cyanobacteria take shelter. These cyanobacteria are referred to as endolithic (literally, inside rocks) (plate 59).

Shores with sedimentary rocks (e.g., sandstone and limestone) may harbor cyanobacteria, including some endolithic forms that dwell in crevices created by the dissolution of calcium carbonate (plate 60). There may be lots of seaweed on the rocks as well. Try to find places where the seaweeds are not dominating the photic zone, for example, just above the barnacles in the area sometimes populated by small snails. There are several other organisms in this zone, however, that can appear to be cyanobacteria. These include the crusty black marine lichen *Verrucaria* and the brown alga *Ralfsia*, which forms tarlike spots. Cyanobacteria such as *Calothrix* may be present as a slippery brown-black, stain-like coating on the rock, with a texture similar to a fine, short velvet. *Lichina*, a cyanolichen may also be present in this zone (see chapter 14).

MUDFLATS AND SALT MARSHES

Mudflats and salt marshes are wonderful places to see bacterial communities in action. In shallow waters, there may be dark, velvety layers of cyanobacteria and floating or attached globular colonies of *Nostoc*, up to a couple of centimeters in diameter (see plate 58). Black, slimy clumps of matted filaments growing on eel grass and seaweed might be *Lyngbya*, a cyanobacterium that can also form felt-like mats (plate 61). In this context it is sometimes called "mermaid's hair." However, there are many filamentous, *eukaryotic* seaweeds that can also look like mermaid's hair. To distinguish the two, pull apart a tiny section to see the actual filaments. Cyanobacterial filaments are nearly microscopic, closer in size to lint than to thread.

If you can smell sulfur in the air when you stir up the mud, then you may be able to see layers of cyanobacteria on top of layers of purple sulfur bacteria and deep, black mud full of heterotrophic sulfate reducers (see chapter 9). If the cyanobacteria are subject to much drying, they may be brownish or reddish. These are good places to collect sediment for making a Winogradsky column (see appendix A). If the mudflat sediments are especially stable (i.e., not subject to waves or foot traffic), you may even find cyanobacterial mats.

In salt marshes, look for places where sediments are relatively stable (not flooded by daily tides) and where waters are somewhat stagnant, perhaps subject to evaporation and increased salinity. There may be a sulfur smell in the air. Look closely at the sediments for a greenish-black coloration at the top. This layer is likely to be photosynthesizing cyanobacteria. If you gently brush or scrape away a thin layer of green, you may find a pink-purple layer directly beneath, which consists of photosynthetic purple sulfur bacteria (see chapter 9). Below this, the sediments will probably look dark brownish-black from the activities of sulfate reducers, which produce hydrogen sulfide as a waste product and give a distinctive smell to the environment. There may also be web-like strands or even a film of white in places where the water is especially stagnant (and smells of hydrogen sulfide). These are colorless sulfur bacteria such as *Beggiatoa* (e.g., plate 39).

MICROBIAL MAT COMMUNITIES

Some hypersaline or very sulfur-rich environments near the seashore are home to well-established communities of cyanobacteria that do more than just color the sediments. Such communities sometimes form mats several millimeters thick consisting of interwoven filaments of cyanobacteria. The gel-like outer coverings, or sheaths, of the filaments give a shiny gelatinous consistency to the mat if it is wet. Dry mats can be revived easily by wetting, and the wet mats are also surprisingly sturdy and fabric-like (see plate 61).

Note that microbial mat communities may be outcompeted by eukaryotic algae and plants, and they are much more subject to grazing and to destruction by foot traffic or tidal waters. Therefore, look for microbial mats in extreme seashore environments where plants and algae are not thriving and where sediments are not regularly disturbed. Lagoons separated by dunes or shallow coves are likely places.

Microbial mat communities usually have the same layering of microbes that is characteristic of sulfur-rich environments. Beneath the blue-green mats lies a pink-purple layer, the purple sulfur photosynthesizers, and below that the black, sulfur sediments of the sulfate reducers. Other sulfur bacteria, such as *Beggiatoa*, may also be found. For a more complete picture of the entire community, see chapter 9.

A compilation of classification systems for cyanobacterial mats is given in table 13.1. All types of mats are easily visible with the naked eye and are among the most distinctive of bacterial structures (plates 62–65). They may be thought of as architectural features of sediment, especially if they become encrusted and hardened with minerals. Cohesive, layered mats are called stromatolites (layered stone); the mats that actually lithify (become rock-like) become part of the fossil record. Today, there are relatively few places where lithified stromatolites are forming. The requirements include lots of dissolved calcium carbonate and relatively few eukaryotic herbivores. A famous (although difficult to access) area is Shark Bay, Australia. Stocking Island (Exuma Cays) in the Bahamas and Chetumal Bay in Belize also have lithifying stromatolites (plate 66). Some freshwater lakes that are rich in calcium carbonate may also form stromatolites; examples may be found in Green Lakes State Park and Onondaga Lake, both in New York State. Note that one feature of cyanobacterial versatility is their ability both to dissolve calcium carbonate (as in the endoliths) and to deposit calcium carbonate (as in the lithifiers). Fossilized stromatolites are discussed later in this chapter.

VIEWING FRESHWATER AND MARINE CYANOBACTERIA UNDER A MICROSCOPE

Using a hand lens, you should be able to see the filaments or coccoids of many cyanobacteria. Even with your naked eye, in fact, you can see the lint-like filaments that make up a microbial mat. Gently tear one apart; it's almost like tearing a bit of felt and seeing the ragged edge of fibers.

If you have a microscope at 400× or 1,000×, you might be able to distinguish some of the main groups of cyanobacteria and to sort them out from eukaryotic algae (plates 67–69). As with any wet mount, take care to remove every grain of sand either by picking

TABLE 13.1. Types of Cyanobacterial Mats

Name and composition*	Moisture	Degree of lithification (encrustation and solidification with minerals)
Laminated biscuit (soft, rounded, gelatinous, 2–5 cm)	Subtidal (submerged)	—
Stromatolite (rounded, submerged, layered communities hardened or lithified with minerals)	Subtidal (submerged, unless fossilized)	Encrusted and solidified
Mamillate or pustular mat (gelatinous with bumpy, wart-like texture)	Intertidal	—
High-pinnacle mat (up to 1–2 cm high cone-like tufts sometimes arranged in concentric rings ["fairy rings"])	Intertidal	—
Low flat (or smooth) mat (coherent, smooth, leathery or felt-like)	Intertidal	—
Farbstreifensandwatt (color-striped sand flats with very subtle texture)	Intertidal	—

Convoluted or blister mat (coherent, leathery-folded, wrinkled, buckled up due to gas bubbles trapped beneath; also called "pee tees")	Rare or seasonal flooding	Sometimes encrusted with minerals
"Tee pees" (cracked, buckled, hard sediments thrust up and propped against each other, forming tent-like structures)	Rare or seasonal flooding	Cemented, often with calcareous minerals
Polygons (dried mats that have cracked into polygonal, often hexagonal, forms—the edges sometimes form tee pees)	Rare or seasonal flooding	Dry and often encrusted or cemented with minerals
Heavenly paper or meteor paper (dry, shriveled bits of folded mat sometimes scattered by the wind)	Rare of seasonal flooding	Leathery or paper-like but very encrusted
Tinten Streif (ink-striped cyanobacterial surfaces)	Rare, seasonal, or localized wetting	Blue-green-black striping on rock

* Descriptive terminology coined by several researchers including Stjepko Golubic, Wolfgang Krumbein, Gisela Gerdes, and Robert Horodyski. Cyanobacterial mats are complex communities, and other types of bacteria play a role in forming mat-like and lithified structures.

them out with fine tweezers or by gently swishing your tiny lint-like sample (gripped firmly with tweezer tips) in clear water. It is worth the trouble, as even one grain will shatter your coverslip when you apply it. Cyanobacterial filaments are 1 to a few μm wide, and the coccoid cells are just a few μm in diameter. Eukaryotic algae are larger by an order of magnitude or more (plate 70).

At present, biologists distinguish five major groups of cyanobacteria. (The classification of Chroococcales and Oscillatoriales is still being rearranged somewhat.) All of these look blue-green in color due to chlorophylls a and b, unless they are dead and decomposing or showing their yellowish or reddish accessory pigments. The three filamentous groups are the easiest to identify under a microscope.

- Chroococcales: single-celled coccoid or rod-shaped, sometimes in aggregates, sometimes in a layer of slime or a sheath (e.g., *Gloeocapsa*, *Synechococcus*) (plate 71).
- Pleurocapsales: single coccoid cells or aggregates of single cells (e.g., *Pleurocapsa*).
- Oscillatoriales: long filaments of tiny identical cells (e.g., *Lyngbya*, *Phormidium*, *Plectonema*, *Microcoleus*, *Oscillatoria*, *Spirulina*) (plates 72–75).

  The subgroup of Oscillatoriales that includes *Lyngbya*, *Phormidium*, and *Plectonema* has been nicknamed LPP because of the difficulties in distinguishing one plain filament from the other—a takeoff of LBB, used by birders for "little brown birds" that are difficult to distinguish in the field.
- Nostocales: long filaments of cells (sometimes looking like strings of beads). Not all the cells look alike; a few are larger and are capable of nitrogen fixation (e.g., *Anabaena*, *Calothrix*, *Nostoc*, *Scytonema*) (plates 76, 77).
- Stigonematales: long filaments of cells. Not all the cells look alike; filaments sometimes form branches (e.g., *Fischerella*, *Stigonema*) (plate 78).

CULTURING FRESHWATER AND MARINE CYANOBACTERIA

Try making a Winogradsky column (see appendix A). Or you may already be "culturing" cyanobacteria if you have an outdoor fountain or dripping faucet with shadowy, damp places—ideal environments. Many aquarium systems that have an "algae problem" may be culturing cyanobacteria. (The problem could be caused by eukaryotic algae too; to tell the difference, you may need a microscope). In a marine aquarium, invasive cyanobacteria might appear as a slimy blue-green or a reddish-purple carpet (the latter called "red slime"). If you don't mind a natural-looking, algae-filled aquarium, you could try inoculating one with cyanobacteria that you have found in the wild. Be forewarned, however: most aquarium fish won't tolerate an overgrowth of photosynthesizers. If the algae or cyanobacteria become prolific, dead ones will accumulate; then aerobic decomposers will scavenge all the oxygen, providing a perfect environment for anaerobic decomposers that produce all sorts of smelly waste products, including hydrogen sulfide. Furthermore, some cyanobacteria produce toxins. A Winogradsky column is by far the easier way to cultivate cyanobacteria.

FIELD MARKS AND HABITATS
OF TERRESTRIAL CYANOBACTERIA

Any moist or intermittently moist soil, rock, tree, wall, or similar substrate has a good possibility of supporting cyanobacteria. In areas where eukaryotic photosynthesizers are dominant, the cyanobacteria are not visible to the naked eye; inhospitable environments are more promising. The more difficult a habitat is for eukaryotes, the better it is for cyanobacteria. Keep in mind that most terrestrial habitats are in a continuum with aquatic habitats, so it may help to refer to the sections on aquatic habitats as well. Look for terrestrial cyanobacteria according to the following guidelines:

1. Sporadic moisture is often fine for cyanobacteria, as they can be quite resistant to desiccation. Look for them on sand dunes, in deserts, and even on bare rocks.
2. Low light is sufficient for cyanobacteria—and is even desirable if eukaryotic algae are to be excluded. The more moisture in an environment, the lower the light should be if you want to find cyanobacteria.
3. As in fresh and marine waters, extreme conditions are more

appealing to cyanobacteria than to eukaryotes. Therefore, look for acidic, alkaline, salty, hot, or cold soils and substrates.

ROCKS AND CLIFFS

Calcareous rocks, especially those that are periodically dampened, may be blue-green-black with cyanobacteria, including endoliths residing under the rock surface. They may be occupying preexisting crevices, or they may be creating their own by producing rock-dissolving acids. Some pebbly beaches of calcareous rock harbor endolithic cyanobacteria, particularly in areas left periodically dry by tides. The calcareous soils of southern Florida, remnants of an ancient coral reef, often have cyanobacterial crusts and endolithic cyanobacteria in the pebbly substrate.

Cliff faces can be fascinating cyanobacterial environments, though sometimes accessible only to climbers. If a trickle of moisture is available as a seeping spring or a splash zone of a stream, cyanobacteria may be present (plate 79). Dark crevices that hold moisture are also promising. Seemingly dry and barren surfaces may green up after rain or during moist seasons. Unlike other types of cyanobacteria, endolithic cyanobacteria often display their blue-green colors rather than the more usual blacks and browns. If things look a bit too green and luxurious, however, the source is probably eukaryotic.

Cyanobacteria often "lead the way" in the ecological succession of dry bare ground or rock. Therefore, plants clinging to the side of a cliff may well have been preceded by cyanobacteria—which may still be there, fixing nitrogen in an otherwise nutrient-poor environment.

PAVEMENTS, BUILDINGS, WORKS OF ART, AND CAVES

Objects made of marble, limestone, cement, or plaster (gypsum)— including statuary and walls (interior and exterior) of buildings— are susceptible to habitation by microorganisms, especially if there is some dampness to encourage their growth. Shadowy walkways and the dark sides of buildings are particularly promising locations for cyanobacterial growth. A wet, shaded fountain provides a wonderful substrate for cyanobacteria. A pale blue-green color is likely to be cyanobacteria, whereas brighter greens in more brightly lit areas are probably eukaryotic green algae. Dried cyanobacteria may form brownish-black crusts or stains, but note that some black crusts can be made of fungi or moss (plates 80–82).

Cyanobacteria and other microbes can be a particular problem when they grow on works of art. The identification of microorganisms in or on a work of art is actually a specialty of art conservation. Art curators often go to great trouble and expense to keep works of art microbe free. The invading microbes may include fungi, actinomycetes (see chapter 11), and various nitrifying bacteria (see chapter 7). If there is even a little light, such as the ambient light within a cathedral, there is an opportunity for photosynthesizers to grow. Among the easier bacteria to identify are the cyanobacteria. Look for dimly lit, moist, or transiently moist surfaces. If the damp wall is made of gypsum-based plaster, the sulfate in the gypsum may be incorporated by the cyanobacteria as a black crust.

The walls of caves are often composed of limestone and are frequently damp and dripping. The only missing key ingredient—for cyanobacteria—is light. Commercially operated caves, however, supply that too, often in the form of dim light bulbs positioned throughout the cave. As you take a cave tour, look carefully wherever you see the combination of a light bulb and a damp wall (plate 83). Any blue-green-black substance is likely to be cyanobacteria. Caves and artwork sometimes become juxtaposed, as in the famous cave paintings of Lascaux, France, and other sites of early humans. Once these caves are exposed to air and light, they become susceptible to microbial growth—and thus destruction of the cave paintings. It is a particularly challenging curatorial problem to keep some of the work available for viewing while also protecting it from damaging microbes.

VIEWING CYANOBACTERIA OF BUILDINGS, ARTWORKS, AND CAVES UNDER A MICROSCOPE

Do not take scrapings from the walls of cathedrals or other works of art. However, when it comes to sampling thick growth on an old fountain, use your judgment; you may be able to take a small sample and examine it for tiny blue-green cyanobacteria, either coccoids or filaments. Determining whether a dry, black crust is fungal or cyanobacterial can sometimes be accomplished by wetting it; cyanobacteria usually become blue-green and gelatinous on wetting. Note that you are unlikely to be able to take samples in commercially operated caves. It is better not to get off the path to do so.

DUNES AND DESERTS

Loose, dry, blowing sand is not a hospitable environment for any organism. However, a little stabilization can produce remarkable results, allowing hardy desert or dune plants to set roots and obtain enough moisture and nutrients to become established. The initial stabilizers of dunes and deserts (both hot and cold) are often cyanobacteria; the crusts they form are referred to by the old-fashioned terms "cryptogamic" or "cryptobiotic." Linnaeus coined the term *cryptogamic* for "plants" (actually algae) that have cryptic or hidden sexual organs. On dry sand, the crust is identifiable by its black to bleached-out blue-green color, often in the vicinity of lichens and plants such as mosses. Green eukaryotic algae may also be in the crust. If you break away a little loose sand, the color becomes more visible. The crust is a firm surface several millimeters thick that keeps the sand from blowing easily. If you walk on it, your footprints will show where you have broken through. Even old footprints or tire tracks continue to be visible and not blown over or obliterated because of the cohesiveness of the cyanobacterial crust (plates 84, 85).

If you can lift up a piece of crust, you'll realize how fragile it is (it quickly crumbles to individual sand grains if you are rough with it) and yet at the same time how stable—forming a secure surface over vast areas of sand. If there are lichens or mosses growing, look under and between them for cyanobacterial crust. You may find communities of large, crusty black lichens and cyanobacteria. The crust not only holds the sandy substrate but it can also hold and retain moisture. If you wet the crust, it quickly comes to life—becoming greener, swelling up with water, and quickly commencing photosynthesis. Moss and lichen spores germinate better on such a moisture-retaining crust and, together with the cyanobacteria, can form the base for a sturdy community of plants. The low areas between dunes (called swales) often have diverse plant communities because water collects in these areas, allowing the cyanobacteria to thrive.

Cyanobacteria also provide nutrients in the form of fixed

nitrogen—an absolutely essential substance in a desolate, sandy environment. They fertilize the sand and make it possible for other photosynthesizing mosses, lichens, and vascular plants to live there.

Tundra regions are desert-like in their sparsity of high vegetation and in their lack of available water when frozen. A surface layer of cyanobacteria as well as lichens and mosses is characteristic of tundra. When tundra thaws it becomes boggy, and in such conditions, which are often nutrient-poor, cyanobacteria thrive.

Bare tropical soils can also be like deserts if vegetation has been burned or cut away, leaving nutrient-poor hardpan. In some cases, if moisture is sufficient, reddish-purple mats (or perhaps just dry crusts) of *Porphyrosiphon*, a cyanobacterium, may be found.

Even a well-trodden path may have desert-like qualities, from a microbial point of view. Although there may be an apparent lack of life on the hard-packed soil, a closer examination may reveal the dark pigments of a thin layer of cyanobacteria, capable of becoming more colorful after a rainstorm.

VIEWING CYANOBACTERIA OF DUNES AND DESERTS UNDER A MICROSCOPE

Extract a little piece of moistened crust and place it in a drop of water on a slide. Mince it up well, and remove every bit of sand. This is the difficult part—but essential, since even one grain of sand will spoil the preparation by cracking your coverslip. You may want to flood the slide with water and then gently remove the tiny lint-like flecks of green. These can then be placed onto a fresh slide. Check carefully for any sand grains before applying the coverslip.

Once you can view your prepared slide, you should see lovely green filamentous cyanobacteria such as *Oscillatoria*. *Nostoc* and related genera may look like strings of beads. You may also see filamentous green algae, which are longer and thicker than the cyanobacteria. Some cyanobacterial cells may look dark (even black) and lifeless, but most should look like healthy photosynthesizers. There is actually an advantage to having dead filaments among the living ones because the former still retain water and, since they are dark, serve, as "sunglasses" for the rest of the living community (too much sunlight can be damaging to the photosynthetic pigments). In addition to photosynthesizers, you may find rotifers, which are tiny animals that have wheel-like mouth parts. They are highly resistant to desiccation.

CULTURING CYANOBACTERIA OF DUNES AND DESERTS

Note that in some desert and dune conservation areas, you might be requested not to disturb the crusts, as they provide nutrients and stability to fragile communities. If you are allowed to sample, collect a bit of crust, submerge it in fresh water (simulating rain), and place it in the sun. It should quickly become greener and form bubbles of oxygen, a sign of photosynthesis. You may also notice a slight swelling of the crust, as the organisms absorb water. This is a temporary culture, however, since it is difficult to simulate conditions that support only crustal cyanobacteria. After a few days, other photosynthesizers, including eukaryotic algae, will take advantage of the sunny, aqueous conditions, possibly outcompeting the cyanobacteria. You would have to simulate periodic drying to maintain only the crustal-type cyanobacteria.

GLACIERS

Windswept, icy glaciers may seem to be lifeless environments. On closer examination, however, especially by microbiologists, it becomes apparent that wherever there is a little liquid water, there is the possibility of finding organisms. Cryoconite holes (also called sun cups or algal pits) are a case in point. They begin as tiny pinpoints where a bit of dark dust or debris has fallen on the ice. Because dark colors absorb light and heat more readily than white colors do, there is the possibility of a little melting under and around the particle. As the hole melts into the ice, more debris accumulates, eventually forming a significant hole, centimeters in diameter and filled with liquid water (fig. 13.1). Because of the angle at which sunlight slants into the hole, the circumference is often not circular but D-shaped. If the melting is substantial, larger pools may even form. (Note that a large chunk of dirt may have a slightly different effect from that of cryoconite holes, insulating the areas beneath such that when the surrounding area melts, the dirt is left on top as a sort of pedestal or "dirt cone.")

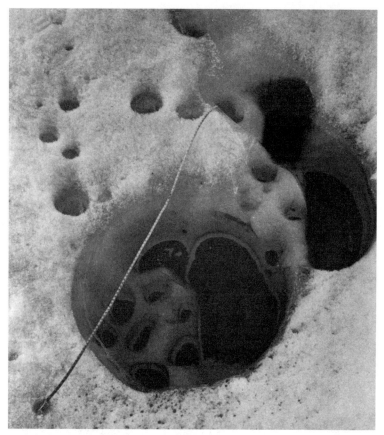

FIGURE 13.1. Cyanobacteria forming D-shaped holes in a glacier in Greenland. (Photograph from United States Army Corps of Engineers, Research Report #50, 1958.)

Microorganisms blown in on the wind colonize these cryoconite holes. Cyanobacteria in particular can form photosynthetic communities in them, and eukaryotic green algae may also be present. The dark pigments of the cyanobacteria enhance the growth of the holes such that they become even more protective for the microbes within. Once photosynthesizers are established, an assemblage of heterotrophic microorganisms can thrive, yielding complex, aquatic communities of organisms such as tardigrades, rotifers, and even tiny "ice worms."

Considerable research has been conducted on cryoconite communities on Arctic and Antarctic glaciers—which are most likely inaccessible to users of this guide. However, Glacier National Park in Montana has several glaciers, and Alaska and Canada are home to many. *Note:* Hiking over glaciers can be very dangerous; it is assumed that the reader will take precautions before venturing out onto one. The *Hiker's Guide to Glacier National Park* by D. and S. Nelson advises never to go onto a glacier without an experienced guide and to be aware of the potentially lethal hazards of hypothermia, deep crevices, and variable weather. If you are exploring parts of a glacier with a guide, look for melted holes that are centimeters (1–10 or more) deep, of various diameters (millimeters to meters) often D-shaped, and with dark, watery bottoms. In the spring and early summer, look especially at the melting edges of the glacier. If you are allowed to take samples, note the cohesive, matlike consistency of the community and the dark black-blue-green pigmentation.

Other things that you might notice include a pinkish-reddish tint on the snow or ice, which is a field mark of the eukaryotic green alga *Chlamydomonas*. This species produces a red pigment that shields it from bright sunlight. Colorful stained snow like this may be found persisting in snow areas during the late spring or early summer. Various heterotrophic bacteria may also be associated with these temporary snow communities.

The cold, bare soils in the vicinity of glaciers and elsewhere in cold latitudes (such as tundras) may have a crust or layer of cyanobacteria similar to those in water-limited habitats (see the earlier section on cyanobacteria of dunes and deserts). Extensive cyanobacterial mats have also been found in Arctic and Antarctic ice-covered lakes. Furthermore, cyanobacteria and other photosynthesizers have been found in, under, and embedded within sea ice. However, exploration of these habitats is not typical for amateur naturalists.

VIEWING CYANOBACTERIA OF GLACIERS UNDER A MICROSCOPE

If you can view a sample from a cryoconite hole, you may see filamentous cyanobacteria as well as eukaryotic algae such as diatoms, desmids, and green algae. Heterotrophs may include the multicellular (yet tiny) desiccation-resistant tardigrades and rotifers. Insects and nematode worms may also be present if the community is well established.

CULTURING CYANOBACTERIA OF GLACIERS

To simulate glacial conditions, you would have to set up a sort of combination refrigeration and light system and monitor it obsessively. It is a project beyond the scope of most amateur naturalists.

FOSSIL CYANOBACTERIA

Bacteria of all kinds have flourished on the Earth since the origin of life about 4 billion years ago. Until about $2\frac{1}{2}$ billion years ago, in fact, cyanobacteria covered every possible moist place and even most dry places that received periodic rainfall or flooding. In some areas, these ancient bacteria became fossilized and are observable as microfossils or stromatolites; cyanobacteria also played a part in creating fossil fuels, "death masks," and some metal deposits.

MICROFOSSILS

Bacterial cells are tiny and delicate and do not leave good fossils, necessitating in a rather elaborate technique for viewing microfossils. However, micropaleontologists (paleontologists who study microscopic fossils, or microfossils) have succeeded in finding and studying microfossils that date back as far as 3.7 billion years. A brief description of the technique used is given here.

First, the micropaleontologist finds a promising sedimentary rock, preferably one that is not composed of carbonates (e.g., limestone) because carbonate crystals are large and tend to crush tiny cells during the fossilization process. Silica-based rocks are favored because the crystals are often tiny and the resulting rock (e.g., a chert, which is a glassy silicon dioxide) is a hard, glassy, protective environment for fossils. Using a diamond-edged saw, the micropaleontologist cuts very thin slices of the rock and mounts them onto microscope slides. Then comes the tedious work of searching through all the sections; not all cherts have microfossils, not all microfossils are well preserved, and those that are often look like nothing more than tiny empty spheres and filaments.

One good hypothesis for identifying these remains is that the microfossils are mostly (or entirely) photosynthesizers—and perhaps cyanobacteria in particular. In general, photosynthesizers are outstanding producers of large, tough, carbon-based molecules such as cellulose. Think of the weighty, solid biomass of trees. Microbial mats, too, have a certain weighty solidity to them. The cell walls of cyanobacteria are thick and often have additional layers, or sheaths, of gelatinous material; such structures fossilize well. Other photosynthetic bacteria (e.g., *Chloroflexus* and related green sulfurs) are also candidates for the identity of these fossils, but cyanobacteria continue to be a good guess.

Some of the most successful ancient cyanobacteria produced highly visible structures that can be seen today; these are sedimentary rocks called stromatolites. At least two areas within the United States and several in Canada have ancient stromatolites that are reasonably accessible.

ANCIENT STROMATOLITES

At their greatest abundance, cyanobacteria built huge reef and mound communities in shallow sunlit waters; the ocean floor in some areas must have been covered with them. Stromatolites—the lithified remains of layered cyanobacterial communities—look like cyanobacterial "skyscrapers," standing centimeters to meters tall.

Stromatolites form when cyanobacterial mats or felts form a cohesive layer on the surface of a sediment in shallow water. Similar mats are still formed today in stable fresh, salt, and thermal waters (e.g., see plates 62a–c). When fresh sediments cover the mat, they block sunlight. Cyanobacteria and other bacteria continually maneuver for position, gliding up out of the sediment and back into the gelatinous materials, thereby forming a layer of trapped sediment beneath them. In very stable environments, many such layers might form. However, any sort of wave turbulence or consumption of the mat material by heterotrophs tends to limit the thickness of the mat and the number of layers.

On ancient Earth, some cyanobacteria must have lived in very calm and eukaryote-free waters, because they formed communities hundreds of layers thick, looming up as gigantic solid domes of trapped sediment. Over the course of millions of years, some of these stromatolites became completely lithified either with chert or calcium carbonate. These fossils may be found in certain places where ancient rocks lie exposed. The oldest stromatolites (about 3½ billion years old) are in Australia and South Africa. Stromatolites that formed about a billion years later may be seen in parts of Canada where the ancient Canadian shield (tectonic plate) lies exposed. It would be too specialized a digression in geology for this guide to describe the major stromatolitic sites in North America. However, four areas in particular are reasonably accessible to amateur microbiologists interested in the geological history of cyanobacteria.

Sites for Viewing Ancient Stromatolites

Yellowknife Area in Northwest Territories, Canada The entire Northwest Territories of Canada has a wealth of ancient stromatolitic sites, some with cyanobacterial mounds of spectacular size that are well known to geologists. Yellowknife, the capital of the Northwest Territories, would be a good starting point for a geological expedition. Visits to nearby sites can be arranged through professional tour guides in the area. The Northwest Territories contain hot springs and glaciers that are also worthwhile destinations for viewing bacteria.

Thunderbay and Gunflint Areas of Ontario, Canada On the northern banks of Lake Superior are stromatolites where the first microfossils were discovered, by Harvard paleontologist Elso Barghoorn. Kakabeka Falls Provincial Park contains stromatolites that are about 2 million years old, though others are more recent. Look at local brochures and signage to determine the age of the rocks you are viewing. Banded iron formations (described later) are also visible in this area.

Glacier National Park, Montana/Waterton Lakes National Park, Alberta These parks are worth visiting for many reasons, including the extensive wilderness and the glaciers. Siyeh Mountain in particular has exposed stromatolites that are about 1.3 billion years old. The Siyeh area is accessed by hiking trails off the main road, and

Figure 13.2. Stromatolites in Ballston Spa, New York.

the best views are above the timberline. The stromatolites of this area have been described by geologists as being among the most diverse in shape and size.

Petrified Sea Gardens, Ballston Spa, New York   The impressive assemblage of stromatolites at Petrified Sea Gardens (figs. 13.2, 13.3), is highly accessible even for young children, as are stromatolites in this area, such as in Lester Park. The stromatolites are fairly recent, about 500 million years old, which is right on the cusp of their almost complete disappearance in the fossil record due to grazing by and competition from eukaryotes. This site is highly recommended,

FIGURE 13.3.   Stromatolite sample from Ballston Spa, New York.

especially if you are eager to see large stromatolites but are unlikely to travel to remote areas that require serious hiking.

Also keep in mind that stromatolites may be viewed in certain museums that specialize in ancient fossils. For example, the Paleontological Research Institution in Ithaca, New York, has an extensive collection of stromatolites, some of which are on display. If you call ahead to inquire at such a museum, be sure to specify *Precambrian* stromatolites—more recent ones are not as likely to be of bacterial origin.

CYANOBACTERIA AND FOSSIL FUELS

Some fossil fuel deposits of sufficient antiquity are the product of cyanobacteria. More recent deposits are likely to be composed mainly of eukaryotes, though cyanobacteria might still predominate in extreme hypersaline environments, some of which are associated with salt deposits (salt domes).

It is perhaps ironic that in addition to having the potential to become fossil fuel, cyanobacterial mats are also used effectively to clean up fossil fuel spills. Such a process occurred on a Middle

Eastern sabkha (area of saline plains and lagoons) where cyanobacterial mats flourished naturally and were further enhanced after an oil spill, creating a moist, nutrient-rich environment favorable to oil-degrading heterotrophs.

DEATH MASKS AND KILLER MATS

The cohesive, fabric-like structure of cyanobacterial mats is responsible for an intriguing phenomenon observable in the fossil record. Some animals that apparently died in or on these mats were only partly digested, were overgrown by the mats, and were mineralized with the help of iron bacteria. Such conditions produce distinctive "death mask" fossils, in which soft tissues are preserved as an impression and sometimes enhanced with a rusty color (iron oxides). Some Brazilian fossils of the Cretaceous (65 to 145 million years ago) that are of this type are popular among collectors. The famous Ediacaran fauna of Australia (from about 580 million years ago) are assemblages of soft-bodied organisms that were similarly preserved under cyanobacterial mats.

The German microbiologist Gisela Gerdes has noted in modern cyanobacterial communities an even more extreme version of this phenomenon, which she calls "killer mats." These mats appear to grow over the burrows of various invertebrate animals, trapping and suffocating the occupants. Anaerobic conditions under the mats prevent full decomposition of the dead animals, and thus they leave entombed traces of themselves.

IRON DEPOSITS: BANDED IRON FORMATIONS AND RED BEDS

About 2 billion years ago, the cyclical activities of oxygen-producing cyanobacteria (perhaps in conjunction with certain iron bacteria) produced massive iron deposits. Because these bacteria were only active at certain times, a distinctive banding pattern was produced in the fossil record; these are now called banded iron formations. Whether this banding is the result of daily, light-dark cycles or longer seasonal cycles is not clear. Either way, it appears that reduced iron was present in large quantities in some bodies of water. As a result, wherever oxygen-producing photosynthesis was occurring (during the day? during the summer?), iron was oxidized and deposited, forming red layers in the sediments. At night or in the winter, iron-poor sediments were deposited, forming the gray layers that separate the bands of iron.

Such banded deposits would be much less likely to occur today because we have so much oxygen around day and night, summer and winter. Two billion years ago, however, the amount of oxygen in the air and water must have varied cyclically. Cyanobacteria had already been evolving for about 1½ billion years. It apparently took that long for enough oxygen to build up to cause the deposition of those oxidized sediments. By about 1.7 billion years ago, oxygen seems to have been more evenly distributed. Evidence for this is that most iron deposits of this age and younger are "red beds," made entirely of oxidized iron.

Both the banded iron formations and the red beds comprise the largest iron deposits of economic use. Most items made of iron today, then, were deposited by cyanobacteria 1–2 billion years ago. Therefore, most iron objects can be considered cyanobacterial field marks.

Modern cyanobacteria still oxidize and precipitate whatever reduced iron is in their vicinity. In most cases, however, the deposits are less dramatic than the ancient ones just described. With so much oxygen around, rusty (oxidized) iron deposits can form almost anywhere, and cyanobacteria are usually no longer the pinpoint source of oxidations. (See chapter 7 for a discussion of other iron-oxidizing bacteria.) Sometimes an oxidized iron layer can form in the layers of a complex mat community.

OTHER METAL DEPOSITS

Gold, uranium, and other heavy metals are not often found in solution, because they are relatively scarce in the Earth's crust and are quite stable as solid metals in an oxygen-rich environment. However, more than 2 billion years ago there must have been many reduced (unoxidized) metals in solution, mobilized from one place to another via rivers, streams, and groundwater. Evidence for this is the age of some of the greatest ore deposits on Earth.

Most of the world's supply of gold is in South Africa in a deposit that covers a huge ancient river delta, now mostly buried deep underground. In appears that when the gold was deposited about 2½ billion years ago, it was in solution or in the form of tiny particles in the waters of several large rivers, all of which converged into a large delta. Cyanobacterial mats probably covered the delta, flourishing in the shallow water and unburdened by competition from eukaryotes. The cyanobacteria likely oxidized and precipitated out whatever gold was in solution and also trapped tiny gold particles in their mats. The result was a massive gold deposit. This deposit was discovered because a later geological event tilted the deposit up to expose a single edge of it at the Earth's surface. The rest is now accessed by deep mine shafts and tunnels.

Because this gold deposit is the largest in the world, and because humans are quite conservative about gold—either keeping it or recycling it into new objects—chances are excellent that the gold in a ring you might be wearing was precipitated by cyanobacteria about 2½ billion years ago. (See chapter 7 for how the gold might have been placed into solution in the first place by *Thiobacillus* bacteria.) Gold deposits of similar age and produced under similar conditions may also be found in Canada and South America.

Geologists hypothesize that deposits of other metals, such as uranium and copper, were formed in the same way. In chapter 7, the practical applications of this phenomenon are described for copper mining.

SUMMARY: FIELD MARKS AND HABITATS OF CYANOBACTERIA OF FRESH WATERS, MARINE WATERS, AND TERRESTRIAL HABITATS

Freshwater habitats (especially low-nutrient, low-light areas with extremes of temperature or pH) (*Note:* cyanobacteria may be black, blue-black, yellow, brown, red, or purple—almost any color but the bright grassy green of eukaryotic algae.)
- brown-black patches (crusty when dry, slimy when wet) above water line in shady areas
- reddish-orange floating gelatinous masses (*Gloeocapsa*)
- globular, mucous colonies (up to 10 cm diameter) floating or attached to vegetation (*Nostoc* colonies can be firm and rubbery; those of *Anabaena* and *Gloeotrichia* are looser)
- yellow-brown films or fuzz, floating or attached to vegetation

- slippery felt-like mats on shoreline rocks or sediments (with lint-sized filaments)
- blue-green coloration on deposits of calcium carbonate
- floating bits that look like tiny particles of chopped grass (*Aphanizomenon*)
- pale tapioca-like floating specks (*Gloeotrichia*)

Wastewater
- nutrient-rich runoff of fertilizer or manure (may be obscured by eukaryotic algae)
- mining operations that use cyanobacteria and algae filters to capture metals from wastewater

Marine habitats (especially low-light, hypersaline, or periodically dry areas unobscured by seaweeds and other eukaryotic algae)
- black or blue-green color on or just under the surface of shells and porous rocks such as sandstones and limestones
- cyanolichens on hard, smooth rocks in splash zones (see chap. 14)
- blue-green-black blooms or layers in salt-marsh or mudflat sediments (see table 13.1 for list of structures that cyanobacteria can form)

Terrestrial habitats (especially low-light, cold, or intermittently wet areas) (*Note:* if composed of cyanobacteria, a dry, dark crust may turn blue-green when moistened)
- cracks and pores of calcareous rocks
- dark, damp sides of walls and buildings
- shaded fountains and damp plaster
- caves (especially commercially operated ones with lights shining on damp walls)
- crusty surfaces of dunes and deserts
- cryoconite holes in glaciers
- microfossils in sedimentary rock and stromatolites (fossilized cyanobacteria; identification is difficult except at known sites)

CHAPTER 14

Cyanobacterial Associations with Other Organisms

Many cyanobacteria are in symbiotic or other relationships with protists, fungi, plants, and animals. The symbioses themselves, along with their characteristics—pigmentations, special structures, and habitats—constitute field marks for the cyanobacteria. In addition to these relationships, we have all around us overwhelming evidence for symbioses that began to be established about 2 billion years ago; some early eukaryotic cells formed associations with cyanobacteria that eventually became so well integrated that the symbionts are now cell organelles—chloroplasts—of algae and plants. Forests, fields, and algae-rich waters all thrive due to the photosynthetic capabilities of the well-integrated cyanobacteria within plant and algal cells.

CYANOBACTERIA AND FUNGI: LICHENS

Lichens are symbiotic associations between eukaryotic green algae (usually *Trebouxia*) or cyanobacteria and fungi (usually an ascomycete, but sometimes a basidiomycete or imperfect fungus). Lichen symbioses are successful because of the ability of the photosynthetic partner to make food and the fungal partner to form a tough, protective covering. This combination allows lichens to thrive on bare rocks, soils, and other dry, nutrient-poor surfaces.

The term *cyanolichen* is considered by some experts to be the proper term for those lichens that contain cyanobacteria (plate 86). The cyanolichens comprise only 8% of all of the approximately 13,500 species of lichens, and about half of the cyanolichens also have green algae partners. The cyanobacteria of lichens can fix nitrogen, which may be an important contribution in environments where nitrogen compounds are limited.

Learning to identify lichens can be a lifetime project for a naturalist, and complete instructions are not within the scope of this field guide. However, lichens with different symbionts can be distinguished. Lichens that contain eukaryotic green algae are often mineral gray or gray-green when dry and a somewhat brighter green when wet. They can also be various shades of yellow. In contrast, cyanolichens are black, brown, or a dull blue-green and are often gelatinous when wet. Table 14.1 shows five common genera of cyanolichens. A strategy for locating cyanolichens might be to focus on the characteristics and habitats of these five genera with the help of a guidebook. (*Lichens of North America* by Brodo, Sharnoff, and Sharnoff is highly recommended.) One of the marine lichens, *Lichina*, is also a cyanolichen; it may be found on some rocky intertidal shores. (For more on this environment, see chapter 13.)

VIEWING LICHENS UNDER A MICROSCOPE

Tear apart a little of the leathery tissue of a cyanolichen. It is likely to make a rather thick preparation under the coverslip, however, so mince the pieces as finely as possible and be sure there are no pieces of grit that will crack the coverslip. You may be rewarded with a view of green cells. Keep in mind that in many cases the morphologies of lichen symbionts are altered. Most cyanobacteria of cyanolichens are filament-forming, but you will not necessarily see their filamentous morphology in your preparation.

CYANOBACTERIA AS SYMBIONTS OF EUKARYOTIC ALGAE

Codium, a large green eukaryotic alga (or seaweed) grows like a weed in many tidal and subtidal zones on the East Coast of North America. The success of *Codium* may be due in part to its nitrogen-fixing cyanobacteria: *Calothrix*, *Anabaena*, and *Phormidium*. Older, thicker branches of *Codium*, which are typically near the holdfast (the base by which it adheres to rocks and shells), may reveal blue-green

TABLE 14.1. Common Cyanolichens

Family and genus	Symbiont(s)	Description
Collemataceae *Collema*	*Nostoc*	"Jelly lichens": papery lichens in rosettes; gray, black, or brown when dry; gelatinous and dark green-black when wet
Peltigeraceae *Peltigera*	*Nostoc* (with some exceptions)	"Toothed lichens": papery (leather-like), brown-gray lichens; rosette, saucer, or fan-like; black when wet; edges ruffled; fruiting structures on tips like teeth or tiny fingernails
Lobariaceae *Lobaria*	*Nostoc* plus eukaryotic alga *Trebouxia*	"Speckled lichens": papery lichens, sometimes warty and pitted
Sticta	*Nostoc* plus eukaryotic alga `Trebouxia*	
Stereocaulaceae *Stereocaulon*	*Scytonema* (sometimes *Nostoc*, *Stigonema*, or other) plus eukaryotic alga *Trebouxia*	"Easter lichens" (some): silver gray, multibranched, coarse, stalked, greenish when wet

clusters or colonies when cut open, although they are likely to be obscured by the green pigment of *Codium* itself.

CYANOBACTERIA AS SYMBIONTS OF PLANTS

BRYOPHYTES

Cyanobacteria of the genus *Nostoc*, a nitrogen fixer, can form associations with some mosses, liverworts, and hornworts—all of which are bryophytes, a group of land plants that lack vascular tissue and typically live in moist environments. These cyanobacteria seem to be transferring fixed nitrogen to their hosts; these symbioses, however, have not been studied as thoroughly as have similar ones that involve nitrogen-fixing *Rhizobium* and legumes (see chapter 6). Presumably, cyanobacteria enable some bryophytes to live in water-logged, boggy, or swampy soils in which decomposition is slow and nitrogen is limiting. And they enable others to live in very arid, nutrient-poor soils, such as sand dunes and desiccated areas.

All plants in a bog have a problem obtaining nitrogen (see plate 55). Decay is slow, and hundreds to thousands of years' worth of dead plants can pile up in the sediments, virtually undecomposed, the nitrogen compounds unreleased. The acidity of bogs acts as a natural antibiotic against many microbial decomposers. Strategies to scavenge nitrogen include insectivory, which is practiced by bog plants such as sundews and pitcher plants (see chapter 12), and symbioses with nitrogen-fixing bacteria. Sphagnum moss, which has cyanobacterial symbionts, is an important photosynthesizer in acidic bogs; it slowly grows into any areas of open water and retains that water in its spongy cells. Acid-tolerant, water-loving plants often use sphagnum moss as a substrate for their own growth. Sphagnum associates with the nitrogen-fixing cyanobacterium *Nostoc.* The cyanobacteria may be either within clear cells of the moss (rarely) or attached to the outside (usually the case). In a sense, the presence of a healthy bog, well filled with sphagnum, is a field mark of *Nostoc* symbionts. Indeed, the very existence of bogs can be attributed in great part to this association, because sphagnum is the most abundant bog plant, and its microbial suppliers of fertilizer (fixed nitrogen) are essential to its survival.

Other genera of mosses have some members associated with a nitrogen-fixing cyanobacterium. These include *Drepanocladus* and *Polytrichum*, both of which are sometimes found in peat bogs. Dry, nutrient-poor substrates such as rock walls, sand, and ashes can harbor mosses such as *Ceratodon, Funaria,* and *Weisia* as well as the hardy *Polytrichum* (plate 87). Under very dry conditions, these mosses can look black and shriveled. On wetting, however, they become green and their symbiotic cyanobacteria are revived and begin fixing nitrogen (see also chapter 13 for a similar phenomenon in desert crust). Urban mosses include *Bryum*, the silvery green, velvet-like moss that grows between sidewalk pavers or in cracks (plate 88). *Ceratodon* can grow on gravel rooftops, looking quite desiccated until it rains. Liverworts such as *Blasia* and hornworts such as *Anthoceros* also have symbiotic cyanobacteria. As with the lichens, a starting place for viewing these mosses is to familiarize yourself with their characteristics and habitats by consulting a field guide to bryophytes.

VIEWING CYANOBACTERIA OF BRYOPHYTES UNDER A MICROSCOPE

Prepare a plant for microscopy by tearing off a small piece and shredding or macerating it on a glass slide in a drop of water. You can use two needles or pins to tease apart the plant tissue. Make as flat a preparation as possible or the tough plant material will break your coverslip.

Try collecting some sphagnum moss growing in a bog and look under the microscope for little beaded, necklace-like arrays of *Nostoc* cells. Larger cells in the "necklace" are heterocysts, which are specialized for nitrogen fixation.

Liverwort species with symbiotic *Nostoc* include *Blasia pusilla*, *Cavicularia densa*, *Marchantia berteroana*, and *Porella navicularis*. Although it is not found consistently in all plants, *Nostoc* may usually be found in dark clumps or spots that are about 0.5 mm in diameter and embedded in mucus on the leaf surfaces. Some species of *Anthoceros* have cavities on the underside of their leaves with *Nostoc* embedded in protective mucus. Others have *Nostoc* on the upper surfaces of leaves.

FERNS AND CYCADS

Azolla, the small water fern, is associated with *Anabaena*, a nitrogen-fixing cyanobacterium. The symbionts are situated in cavities on the underside of the dorsal (topmost) lobe of the compound leaf. The world's rice crop may well depend on this symbiosis, since rice paddies are full of this nitrogen-rich floating fern (plate 89). Cyanobacteria growing separate from *Azolla* are also sometimes cultivated in rice paddies to provide fixed nitrogen as fertilizer.

Cycads are gymnosperms (nonflowering seed plants) that are native to tropical areas but are often grown in gardens and greenhouses. Some species have specialized structures called coralloid roots, which look like short, thickened fingers about 1 cm long. These roots usually lie just below the soil surface, or just above the soil in older plants or greenhouse specimens (plate 90). Cutting through the roots might reveal a blue-green tint of *Nostoc*; this symbiont is fixing nitrogen for the plant. (Heavily fertilized ornamental cycads may not contain *Nostoc*, since fixed nitrogen is readily available.) During the heyday of cycads in the Jurassic and Cretaceous periods, nitrogen-fixing *Nostoc* might have been responsible for enabling cycads to proliferate in huge forests worldwide.

VIEWING CYANOBACTERIA OF FERNS AND CYCADS UNDER A MICROSCOPE

As with the bryophytes, cyanobacteria can be found inside the tissues of ferns. Collect a little *Azolla*, and crush or mince open the dorsal lobe of the compound leaf to release the cyanobacteria. If rice paddies are not available, you might find *Azolla* cultivated in a water garden.

It can be difficult to prepare tough plant tissue from cycads to make it flat enough to observe under a microscope. A hand lens examination of a sliced root might be a more effective way to get a closer look at the blue-green tint of the plant tissue.

BROMELIADS AND *GUNNERA*

Water-filled cavities of plants are potential habitats for cyanobacteria and other microbes. The technical term for such cavities, *phytotelmata*, includes all sorts of waterproof plant structures such as bamboo stems and stump holes. Bromeliads are members of a large tropical plant family, some of which form phytotelmata called tanks—specialized leaves that hold water in cavities at their bases. Depending on the size of the plant, a tank bromeliad can hold milliliters to liters (as much as 20 L) of water. Entire aquatic communities can develop in bromeliad tanks, composed of organisms such as bacteria, algae, insects, and amphibians. Many tank bromeliads are epiphytes, which are specialized plants that live on other plants and do not have roots in contact with the soil. Tank bromeliads seem to have "solved" the problem of having no contact with the soil by taking up minerals from their tank community via specialized cells. Cyanobacteria live in many tank communities, and the nitrogen-fixing ones may well contribute directly or indirectly (via other community members) to the bromeliad host. The cyanobacteria in turn get a relatively stable little body of water in which to grow. For more on bacterial activity in bromeliad tanks, see chapter 12.

Gunnera is a genus of tropical wetland plants that is characterized by its enormous leaves. This is the only genus of flowering plant known to have internal cyanobacterial symbionts. *Nostoc* grows in the tissues at the junction of the stem and roots, just below the soil surface. Cutting into the stem reveals blue-green patches.

VIEWING CYANOBACTERIA OF TROPICAL PLANTS UNDER A MICROSCOPE

You might scrape some blue-green material out of the stem of *Gunnera*. Remove as much tough, fibrous plant material as is possible to make a preparation containing tiny cells of cyanobacteria among the much larger plant cells. You may also want to collect water from tank and other phytotelmata communities to look for cyanobacteria as well as other organisms, including eukaryotes.

CYANOBACTERIA AND ANIMALS

FLAMINGOS AND HUMANS

Flamingos are large pink birds that can be found in some tropical, saline lagoons, usually feeding on bacteria, algae, or crustaceans. The color of their plumage is due to carotenoids, a pigment that derives from the cyanobacteria and algae, or indirectly, crustaceans, in their diets. Flamingos kept in zoos may lose their pink coloration if a supplement of carotenoids is not provided. Different flamingo species specialize in different sizes of food, according to the fineness of the straining mechanism by which they filter and collect organisms. The finest filters collect cyanobacteria. In particular lesser, Andean, and James flamingos are all cyanobacterial feeders.

Their pink feathers, therefore, are a field mark of the pigments of cyanobacteria selectively filtered from the local water. However, any flamingo, no matter what its filter size—and for that matter any pink-colored, water-feeding bird—is at least indirectly indebted to cyanobacteria for its pinkness. Cyanobacterial pigments (along with some algal pigments) become concentrated in the food chain and end up in the flamingos via a diet of crustaceans, which in turn have eaten the bacteria.

Some humans consume cyanobacteria in fairly concentrated quantities. *Nostoc* appears in several Asian cuisines. In parts of China, for example, *Nostoc* is a culinary delicacy called *shi*. Traditional foods of Central and South America also include cyanobacteria. In Chile and Peru, for example, *Nostoc* prepared as food goes by the names *picantes, chupa, locra*, and *mazamrro*. In some Western cuisines, *Spirulina* is considered to be a health food. It is grown, dried, and packaged in capsules for consumption. In a similar process to that of flamingo coloration, humans who ingest an excess of carotenoids from any source, including carrots, can turn orange!

THREE-TOED SLOTHS

Three-toed sloths are remarkable animals—even without their amazing association with cyanobacteria. They are leaf eaters, specializing in cecropia trees. Like all herbivores, they have digestive symbionts, as described in chapter 12, but they also have symbionts in their fur.

The sloths' sedentary lifestyle leaves them vulnerable to predators, but sloths are well camouflaged with multicolored fur. Individual hairs in their coat have ridges along which the cyanobacteria *Trichophilus* and *Cyanoderma* grow, especially during rainy seasons. These cyanobacteria also live in association with other fur-dwelling organisms, including a moth that spends its entire life cycle on the sloth and which may graze on the cyanobacteria. In addition to forming a visual camouflage of yellow to dull greens, this symbiotic fur community may disguise the scent of the sloth as well. Sloth feces are also odorless, perhaps due to the efficiency of its digestive symbionts. Thus, the sloth is well concealed during its many motionless hours. In dry seasons, sloths become brown, apparently due to their dried-out cyanobacteria, and thus they continue to blend in with dried-out foliage (plate 91). Two-toed sloths (*Choloepus*) are related to the three-toed sloths (*Bradypus*) but have different habits and lack the cyanobacterial symbionts of the latter. Both types of sloths are closely related to anteaters and armadillos.

Three-toed sloths (unlike two-toed sloths) have never been successfully maintained in zoos. Something in their very particular diet apparently is missing in a typical zoo diet for leaf-eating herbivores. Wild sloths may be eating ants that live in cecropia trees, or perhaps cyanobacteria and other organisms groomed from their own fur. These intertwined relationships of sloths and their organisms may be difficult or impossible to maintain in captivity.

Are there other fur symbionts of animals? An intriguing report from microbiologist Ralph Lewin stated that filamentous cyanobacteria sometimes establish themselves in the hollow white hairs of polar bears in captivity. However, this seems not to occur in wild polar bears.

ELEPHANTS

Male elephants in "musth" (testosterone-induced readiness for mating) undergo a number of physiological and behavioral changes. In addition to becoming much more aggressive and vocal, producing deep rumbling sounds, they release bodily exudates in their urine and from specialized glands at their temples. Some of these behaviors appear to be signals intended to attract female elephants, who also have distinctive urinary compounds and vocalizations. But where do cyanobacteria come in? The penis of an elephant in musth can become covered in a layer of algae and bacteria, some of which may be cyanobacteria. The fact that this is typical of healthy elephants suggests a symbiosis of some kind. Taking a sample would probably require tranquilizing the animal, a procedure attempted only rarely with elephants. Therefore, elucidation of the presumed symbiosis awaits the right intrepid researcher.

SPONGES

Sponges are invertebrate animals found most commonly in tropical marine waters. They are considered to be similar to some of the first animals to evolve. Sponges are sedentary, usually amorphous, variable in size, and perforated with holes through which they feed on microorganisms in the water. Examination of a natural bath sponge, which is a dried piece of a large sponge animal, will give you a good idea of how amorphous and perforated these animals are. Sponges are often classified according to their "skeletons," which may be made of protein, calcium carbonate, or silica (glass). Proteinaceous and carbonate sponges, when found in shallow tropical waters such as in coral reefs, often have symbiotic cyanobacteria situated on and throughout the sponge, and sometimes even within the sponge cells. These sponges are sometimes referred to as "cyanophilic" (cyanobacteria lovers). All sponges contain bacteria (of many types), some of which are taken in as food and others that seem to be residents. Sponges with more than 20% of their biomass composed of bacteria are called "bacteriosponges" by German researcher Joachim Reitner and others; percentages range up to 60 to 70% of the sponge's mass. Speculations as to the role of cyanobacteria in sponge nutrition include the possibility that the bacteria are fixing nitrogen and transferring it to the sponge, or that they are providing some photosynthetic product (food).

ASCIDIANS

Ascidians (also called sea squirts or tunicates) are intriguing marine invertebrate animals. Often they look like blobs of jelly—or several blobs, if a colonial form. When relaxed, ascidians may show the system of holes by which they filter microscopic food from the seawater. The sedentary ascidians are similar to sponges in their behavior, but while sponges are considered "primitive," ascidians have undergone many evolutionary changes and are actually closely related to chordates (our own animal phylum). Some tropical ascidians of the Pacific have *Prochloron* cyanobacteria as symbionts. In fact, most of the known species of *Prochloron* are obligate symbionts of ascidians. *Prochloron* seems to have undergone evolutionary changes in its pigments, since it contains both chlorophyll a and b but not phycobilin, the typical secondary pigment of cyanobacteria. Some taxonomists believe that this assemblage of pigments is enough to place *Prochloron* in its own group separate from the cyanobacteria, which have only chlorophyll a and phycobilin. DNA sequence analysis, however, suggests that *Prochloron* is at least a close relative of cyanobacteria.

BRYOZOANS

Bryozoans (moss animals) are frequently found attached to a variety of marine surfaces. They are tiny, sessile (stationary) animals, but some can look as green or blue-green as a layer of moss or a cyanobacterial mat. Some have such a concentration of cyanobacterial symbionts that they appear greener than the surrounding area.

ECHIURAN WORMS

Some echiuran worms of the Pacific coral reef and intertidal zones have cyanobacterial symbionts. These worms include *Bonellia fuliginosa* and *Kedosoma gogoshimerse*, both of which have green spots. Other spotted echiurans may have symbionts as well. The green echiuran species *Bonellia viridis* does not have cyanobacteria, however. Its color is due to the green pigment bonellin. Noncyanobacterial pigmentation is discussed at the end of this chapter.

LOBSTERS AND CRABS

A surprisingly complex community dwells in the bronchial chambers of lobster species such as *Homarus vulgaris*. The feathery bronchia of this lobster are filled with cyanobacteria, other types of bacteria, and even a small polychaete worm that feeds on the bacteria. The worm may be a "cleaning symbiont" that keeps the chamber free of an excess of bacteria. Cooked lobsters, the sort with which most people are familiar, do not provide the best view of this community. You would have to perform a somewhat inhumane live dissection in which you removed the white, feather-like appendages from beneath its cephalothorax.

As with many shells in aquatic environments, especially in the photic zone, the borrowed ones of hermit crabs may become covered with various sedentary animals, algae, and cyanobacteria. This is a community that could be sampled (with gentle scraping) and observed under a microscope.

BLUE-GREEN AND GREEN COLORS IN OTHER MARINE INVERTEBRATES

Not all blue-green pigmentation is an indication of cyanobacteria. Many polychaete worms and some echiuran worms have beautiful, sometimes iridescent blue-green colors, none of which are a result of symbioses. Giant clams, which have eukaryotic algal symbionts, also have striking blue patterns of nonsymbiotic pigmentation. Some of these bluish pigments have been found to contain copper, which is a probable source of the color. A rule of thumb might be that if the color is too brilliantly blue, it is probably not produced by cyanobacteria.

Many marine invertebrates, however, have eukaryotic algae as symbionts. Most notable are the corals, the giant clams, and some sea slugs and flatworms. Any shade of green in a marine invertebrate indicates the possibility of an algal symbiont, though it could also be a result of copper.

VIEWING CYANOBACTERIA OF ANIMALS
UNDER A MICROSCOPE

Dissections of cooked animals or animals found dead are not likely to yield cyanobacterial symbionts. If you intend to dissect a freshly killed animal, first acquire a zoology lab manual to become familiar with the anatomy and dissection procedure as well as with humane killing methods for the animal of interest.

SUMMARY: FIELD MARKS AND HABITATS OF
CYANOBACTERIAL ASSOCIATIONS WITH OTHER ORGANISMS

- some lichens that are brown, black, or dull blue-green, and gelatinous when wet (for field identification, look for the genera listed in table 14.1)
- sphagnum moss (symbiont: *Nostoc*) and other mosses in dry nutrient-poor areas (use a field guide to identify genera listed in this chapter)
- small water fern, *Azolla*, often found in rice paddies (symbiont: *Anabaena*)
- "tanks" of bromeliads
- *Codium* seaweed (symbionts: *Calothrix*, *Anabaena*, and *Phormidium*)
- fur of three-toed sloths (symbionts: *Trichophilus*, *Cyanoderma*)
- layer (with algae) covering penises of elephants in musth
- some marine invertebrates with blue-green pigmentation
- sponges in tropical marine waters
- sea squirts (ascidians) of the Pacific Ocean (symbiont: *Prochloron*)
- blue-green bryozoans attached to marine surfaces
- some echiuran worms of Pacific coral reefs
- bronchial chambers of lobsters

CHAPTER 15

Bacteroides, Gliders, and Their Relatives

Bacteria in the genus *Bacteroides* and in the group known as "gliders" are still being sorted and categorized according to their DNA sequences. Because species of *Bacteroides* have the most obvious field marks, the chapter title leads with their name. However, the gliders are also fascinating. All the bacteria discussed in this chapter are rod-shaped gram-negatives—traits which put them in the ranks with most organisms on Earth, the vast majority of which, you will recall, are microscopic. For a more complete discussion of gram-negatives, see the introductions to the chapters on gram-positives and on proteobacteria. To summarize briefly, Gram's stain is one of hundreds of stains used to identify bacteria, but it also happens to delineate some broad taxonomic groups. All of the gram-positives (bacteria which become permanently colored blue by Gram's stain) form one large, diverse group. The archaea are neither positive nor negative. All other bacteria are gram-negative, and the best known are the proteobacteria. Were it not for their significantly different DNA sequences, which place *Bacteroides* and their relatives in a separate group, these organisms would be indistinguishable from the proteobacteria. The rod shape of *Bacteroides* comes in various forms, ranging from straight to curved to spiral; thus, microscopic examination alone does not easily distinguish this genus from many other bacterial groups.

BACTEROIDES

Bacteroides species are your "other" intestinal bacteria, sharing that bountiful niche with the gram-positives (see chapter 12), methanogens (chapter 2), and the hydrogen sulfide–producing delta proteobacteria (chapter 9). *Bacteroides* of intestines eat what their host eats, especially favoring carbohydrates, which they ferment, producing various acids as waste products. Some *Bacteroides* can also metabolize what their hosts cannot: tough cellulose and other difficult-to-digest plant compounds.

Like most inhabitants of intestines, *Bacteroides* are anaerobic or have a tolerance for low oxygen. The group may also be found in deep anoxic mud and sludge, although these two environments are so popular among microbes that it would be difficult for the amateur naturalist, using only field marks, to distinguish *Bacteroides* from all the other bacteria. *Bacteroides* of mammalian digestive systems have more distinctive field marks than do those in other habitats because of their great numbers and well-defined environment; therefore, they are the focus here.

BACTEROIDES OF TEETH AND GUMS

Some *Bacteroides* get a head start on whatever carbohydrates you (or any other animal) might be eating. They situate themselves in the tiny anaerobic crevices between the gums and teeth and between teeth. There, they join an even more abundant type of mouth bacteria, the gram-positives. The more debris, the more plaque, and the more spaces and cracks, the better for these bacteria.

Unfortunately, the acidic waste products of *Bacteroides* and others contribute to the erosion of tooth enamel and decay. *Bacteroides* are one of the reasons we brush and floss. Healthy mouths are full of gram-positives and *Bacteroides*, and no amount of dental hygiene can keep them from their habitat. However, there is a difference between healthy and unhealthy levels of bacteria. The sugary diets of modern humans, along with our penchant for living long and expecting our teeth to last for the duration, means that dental hygiene is required to keep those gram-positives and *Bacteroides* from becoming so settled that they literally rot their habitat, our teeth and gums.

One relative of *Bacteroides*, *Leptotrichia*, is surmised to have been observed by the 17th-century Dutch naturalist Anton van Leeuwenhoek from scrapings between his teeth. A typical procedure by which van Leeuwenhoek cleaned his teeth included rubbing with salt, rinsing with water, cleaning with a toothpick, and scrubbing with a cloth. Nevertheless, on examining a bit of plaque from between his teeth, he "saw, with great wonder, that in the said matter there were many very little living animalcules, very prettily a-moving" (from a contemporary translation). At least five types of bacteria were in this sample, including one suspected to be *Leptotrichia*. Investigations of the plaque in other people's mouths led van Leeuwenhoek to try a sample from a man who had never cleaned his teeth. Among the many swarming creatures were spirochetes (chapter 16), described by van Leeuwenhoek as follows: "The biggest sort (whereof there were a great plenty) bent their body into curves in going forward" (Dobell, *Antony van Leeuwenhoek and His "Little Animals"*).

INTESTINAL *BACTEROIDES*

On the list of the 25 bacterial species most likely to be found in the feces of healthy humans, *Bacteroides* (as a genus) appears first on the list as well as four other times. A *Bacteroides* relative, *Fusobacterium*, is second on the list. Most of the other bacteria are gram-positives. In fact, bacteria make up about three-quarters of the wet weight of feces, about 20% of which is *Bacteroides*, and another 7% *Fusobacterium*. Therefore, you may consider a significant part (about 27%) of the weight of your own feces to be a field mark of *Bacteroides* and its relatives.

Your healthy, functioning large intestine is also a hallmark of your helpful population of *Bacteroides*. In this anaerobic environment, the bacteria are busy attacking carbohydrates that you did not digest yourself and producing acidic waste products. Indeed, the intestines can have a layer as much as an inch thick of bacteria, many of which are *Bacteroides*. The density of intestinal bacteria may be illustrated with the following calculation. Based on estimates by the microbiologist Sherwood Gorbach, a human colon has a capacity of about 1 liter or 1,000 milliliters (or cubic centimeters). If we allow each colonic bacterium a cubic space 1 micrometer on a side (there are 10,000 micrometers per centimeter), we could pack 1,000 trillion bacteria into a cubic liter. It has been estimated that 100 trillion bacteria reside within the colon—just one order of magnitude away from what microbiologists estimate to be the densest possible packing. This means that each little colonic bacterium is only about 5 bacterial lengths away from a neighbor in any direction.

Our bacterial cells outnumber our own cells to the extent that a complete picture of our genome (our total assemblage of DNA) should perhaps include the genome sequences of hundreds of symbionts; the sequence of our own genome does not convey the full story. The microbiologist Dwayne Savage has estimated that every normal human is "composed of over 10^{14} cells, of which only about 10% are animal cells."

It is difficult to know exactly what we would do without our intestinal *Bacteroides* (and gram-positives). Sometimes our microbial community is disrupted by an opportunistic pathogen, resulting in cramps, gas, and diarrhea. You begin to feel better when your *Bacteroides* and gram-positives restore themselves. However, the system is so complex that it is difficult to ascertain exactly which good things are being done by which bacteria. In general, the presence of your community of *Bacteroides* and gram-positives prevents the proliferation of most pathogens. Be thankful for them!

The role of intestinal bacteria has been investigated in experiments conducted on nonhuman mammals. "Germ free," and thus sickly, mice, for example, are maintained at considerable expense in sterile labs with sterile food, air, and water. Using these unhealthy mice, scientists have demonstrated that *Bacteroides* helps other bacteria colonize the intestinal surfaces by modifying those surfaces in ways that make them more hospitable. *Bacteroides* may also modulate the activity of our own intestinal genes, stimulating them to produce particular products. The intestinal layer of *Bacteroides* may also mediate between our immune system and the constant flow of foreign material (food and bacteria).

VIEWING *BACTEROIDES* UNDER A MICROSCOPE

Each gram of fecal material contains about 10^{11} bacteria, about 27% of which are *Bacteroides* or related species. Take a tiny dab of your feces on the tip of a toothpick and place it in a drop of water on a slide. Add a coverslip and observe. Expect your view to be obscured somewhat by undigested fibrous and globular material. Also, expect to see lots of rod and coccoid bacteria, indistinguishable by morphology alone. Dental plaque is also a wonderful source of bacteria, some of which are *Bacteroides*.

CULTURING *BACTEROIDES*

Bacteroides thrive under anaerobic conditions and therefore are not easily cultured by amateurs, except for the easiest method of all: maintaining them in your healthy intestines.

GLIDERS

Gliding bacteria, or gliders, are a diverse group of bacteria capable of an intriguing type of movement. Rather than using flagella (tiny whip-like appendages), gliders smoothly skate along solid surfaces or each other, sometimes leaving a trail of slime. The mechanism by which they glide is not well understood. Gliding motility is found in three groups of autotrophs, including some green nonsulfur bacteria (chapter 1), sulfur oxidizers such as *Beggiatoa* (chapter 9), and filamentous forms of cyanobacteria (chapter 13). The heterotrophic gliders are the myxobacteria (chapter 8) and the gliding relatives of *Bacteroides* (described here). Many of the heterotrophic gliders specialize in consuming large, tough molecules and even whole bacteria by swarming around their food (like a wolf pack, in microbiologist Martin Dworkin's words). They excrete specialized digestive enzymes that break down large molecules to sizes small enough to be absorbed. Their diets include chitonous shells of arthropods, stringy cellulose fibers of plants, and gel-like agar and pectin.

FIELD MARKS OF GLIDERS

Although the gliding relatives of *Bacteroides* do not present overt field marks, they sometimes give themselves away by their color. In damp cellulose-rich material such as paper or cardboard undergoing decomposition, the predominant microbes are likely to be fungi. However, reddish or yellowish stains on the material can be a sign of gliders such as flexibacteria, which range from yellow to orange to red. The pigment responsible for these colors is flexirubin, and one way to test for it—to be sure the color isn't caused by some other bacterial or fungal pigment—is to drop some 10% KOH (potassium hydroxide) onto the material, which should turn the reds to purples or browns. This reaction is reversible; 1% HCl (hydrochloric acid) makes the pigment red again. Gliders' pigments are so striking that the imaginative microbiologist Ralph Lewin once described the color of a tube of cultured "flexis" (his nickname for flexibacteria) as somewhere between peach and apricot—and then illustrated his lecture with a slide of the culture positioned between the two actual fruits.

GLIDERS AND TEXTILES

Gliders (along with other bacteria, and fungi) are used to derive natural fibers such as flax (linen) and hemp from the tough, stringy, fibrous material of the stems of those plants. "Retting" (an ancient word derived from rotting) is the process by which the hard, woody outer layer of stem material is partially rotted away, exposing the useful fibers within. Traditional retting is done either by placing bundles of stems weighted down in a pond or stream for about 10 days or by setting the stems out on the ground and watering them frequently for 4 to 8 weeks. The former is mediated by cellulose-digesting bacteria including gliders and some gram-positives. Sulfate reducers (chapter 9) also thrive. The latter is called "dew retting" and is dominated by fungi rather than bacteria.

After retting of flax is complete, the remains of the decayed material are broken on a flax-brake and "scrutched," or scraped away. The exposed fibers are then "hackled," or combed, on iron spikes in preparation for spinning (on a flax wheel) and weaving. Fibers of hemp are produced in a similar way. Linen is an especially strong textile because it is made up of the part of the plant that could not be successfully attacked and digested by bacteria. Some naturally dyed linen, especially in shades of blue, presents the field marks of at least two different bacterial groups: the cellulose-digesting gliders (and other microbes) are responsible for the production of the linen itself; and some gram-positive bacteria are responsible for the blues of natural dyes such as those derived from indigo, woad, and some lichens (see chapter 11).

Given the way in which flax is processed, one might think that plant fibers for preparing paper could be handled in the same way. Apparently for a brief time in the Middle Ages, they were: bundles of rags were left to ferment before being used to make rag paper. Artisans of handmade papers made from fibers such as grasses and shredded vegetables also use a similar process. They often begin with a vat of odoriferous rotting vegetation as a first step to rendering the material into pulp. Although fermentation is usually not mentioned in instructions for handmade paper, the complex musty, sulfury, cheesy odors of bacterial activity are unmistakable. The sulfur smell could be produced either by fermenters or sulfate reducers. The standard commercial method for producing paper, however, does not use fermentation; fibers are pounded and cooked ("pulped") in alkaline solutions. Some of these solutions are rich in sulfate. The combination of sulfate and cellulose fibers in wastewater makes ideal conditions for sulfate-reducing bacteria, which is why waters polluted by paper mills are often so smelly.

VIEWING GLIDERS UNDER A MICROSCOPE

If you are sure that you have some decaying cellulose-rich material with appropriate pigments (tested with KOH and HCl, as described earlier), you might be able to identify gliders with some confidence. They are long, thin rods often situated along the long fibers of the cellulose. Cellulose fibers should be the largest, most abundant things in your field of vision. By adjusting the light, you may be able to spot the much smaller bacteria. Gliders on a solid surface have been clocked at 100–150 micrometers per minute, which, at their typical length of 10 micrometers, translates to about 10 cell lengths per minute, or a full cell length every 6 seconds. Thus, a glider can be quite speedy by bacterial standards.

CULTURING GLIDERS

If you have made a Winogradsky column for the purpose of culturing a colorful community of bacteria (and you have been strongly urged to do so throughout this book; see appendix A), you are already culturing some cellulose-consuming gliders and other bacteria. They are slowly feasting on the crumpled paper towels you put in as your bottom layer. Compared with the other occupants of your column, however, the gliders remain invisible. In fact, the whole bottom of the column may well be black, obscuring your view of anything but the hydrogen sulfide–producing sulfate-reducing bacteria. Also, many different microbes (including fungi) break down cellulose, so the mere presence of a decomposing paper towel is not a reliable indicator for any particular bacterial groups.

Sometimes gliders can be lured to a strip of paper (or other cellulose-containing substrate) suspended by a clothespin in a jar of pond or marine water. Spots of yellowish-reddish pigment, especially at the air-water interface, suggest the presence of gliders, many of which are aerobic. Other bacteria as well as fungi may be present as well, however, and may grow more quickly than the gliders, which take about 1 to 3 weeks to appear. Try testing the pigment with KOH and HCl. Also, examine some of the fibers under a microscope to see if there are elongated gliders.

Small amounts of hay or straw (3 to 5 pieces) or other dry plant material in a jar of pond water make a good environment for aerobic cellulose digesters such as gliders. In fact, such enrichments (sometimes called hay infusions) have a long tradition of classroom use to provide protists such as ciliates for microscopy. The slow digestion of cellulose by bacteria can help establish a community of bacteria-eating protists and then protist-eating protists as well as small animals like rotifers. To encourage an overgrowth of bacteria, add lots of dried plant material to your jar (plate 92). This will result in an anaerobic zone in the bottom and may encourage bubbling methanogens and fermenters of many kinds. Again, if cellulose-eating gliders are present, they will be near the top of the jar in the aerobic sections.

A parallel phenomenon to a sparse hay infusion can be created in ponds by adding moderate quantities of fermenting plant fiber (such as a little straw) or allowing existing plant material (such as decaying rushes) to remain. Larger amounts of nutrients, as are present in manure- or fertilizer-laden ponds, paper-producing plants, and retting ponds, become anaerobic and smelly from an accumulation of decaying plants and a proliferation of a variety of bacteria. Try some retting experiments in a pond with small bundles of floating hay or straw versus larger bundles weighted down by rocks to see what different microbial communities develop after days to weeks.

SUMMARY: FIELD MARKS AND HABITATS OF *BACTEROIDES*, GLIDERS, AND THEIR RELATIVES

- mammalian feces (*Bacteroides, Fusobacterium*)
- mammalian gums and teeth (*Bacteroides* and others)
- some discolorations of decaying cellulose, such as paper towels (if color is reddish-yellow and becomes red, purple, or brown upon application of potassium hydroxide) (gliders)
- "retting" of flax and hemp to make linen in the traditional manner via bacterial degradation (gliders; note that a diverse community of cellulose-degrading bacteria and fungi is likely to be involved)

CHAPTER 16

Spirochetes

The spirochetes are easily identified under a microscope by their elongated helical or undulate gram-negative cells (for more on gram-negatives, see the introduction to the gram-positives). As a result of this morphology, they are able to corkscrew themselves through dense substances—mud, or intestinal contents, or, in the case of pathogenic spirochetes, the tissue of a plant or animal. The best-known pathogenic species are responsible for Lyme disease and syphilis.

Spirochetes are abundant in the anaerobic regions of animal digestive systems and in both freshwater and marine sediments. They thrive in cellulose-rich environments, as well, where they and other bacteria ferment various carbohydrates. Spirochetes may also be found in dental plaque, along with a myriad of other bacteria (see chapters 12 and 15). Anton van Leeuwenhoek, the 17th-century Dutch microscopist, was the first to see and describe spirochetes, which he identified in a scraping from his own teeth. A scraping of plaque provides a good opportunity for viewing spirochetes under the microscope.

FIELD MARKS AND HABITATS OF SPIROCHETES

There are two reliable but indirect field marks of spirochetes, each of which is interesting in its own right. Both involve spirochetes as symbionts of invertebrate animals—namely, molluscs (table 16.1) and termites. In both organisms, the spirochetes are associated with the digestive system and are involved in breaking down cellulose and other tough food molecules. The first step to finding spirochetes is to seek either of these host organisms; the second step is dissection.

TABLE 16.1. Genera of Molluscs Containing *Cristispira* Spirochetes

Group	Genus
Oysters	*Crassostrea, Ostrea*
Mussels	*Mytilus, Modiola*
Scallops	*Pecten*
Quahog	*Venus*
Sea pen	*Pinna*
Short razor	*Solen*
Ribbed pod	*Siliqua*
Jewel box	*Chama*
Rock cockle	*Chione*
Cumingia	*Cumingia*
Iceland cockle	*Clinocardium*
Boring clam	*Gastrochaena*
File shell	*Lima*
Lyonsia	*Lyonsia*
Macoma	*Macoma*
Panope	*Panope*
Rock borer	*Saxicava*
Piddock	*Zirfaea*

Source: Albert Balows et al., eds., *The Prokaryotes: A Handbook on the Biology of Bacteria* (New York: Springer-Verlag, 1992). *Note on collection:* Molluscs must be collected fresh to find *Cristispira.* Even a few hours on ice in a fish market will lessen the chance of seeing anything. Not all individuals have spirochetes; also, many molluscs have a crystalline style but do not have spirochetes.

CRISTISPIRA OF MOLLUSCS

Cristispira are large, highly motile spirochetes found exclusively in the digestive systems of some invertebrates. At 100 micrometers ($1/_{10}$ millimeter) or more in length, they are among the largest of bacteria. Their type of symbiosis with an invertebrate host seems to be commensal, meaning that there is a benign relationship in which the spirochete shares in the abundance of food in the host's digestive system but neither benefits nor harms the host. The field mark of *Cristispira* is indirect: you may be reasonably assured that a healthy, well-fed mollusc of the right species contains a concentrated culture of them, ready for observation if you have a microscope.

Get a fresh, live quahog, mussel, or oyster from either marine or fresh water. Dissect the mollusc, which you might well be doing anyway if you had just gone quahogging or oystering, except that you would call the procedure "preparing the mollusc for consumption." First, open the shell and take a look at the large, central, bulbous area, the part of the digestive system that is sometimes called the belly. Boldly cut into this tissue, watching carefully for a rather strange structure, the "crystalline style." The style acts as a sort of pestle, grinding food against a hard, mortar-like structure in the digestive system. It is a long, smooth, narrow rod, 1 to 2 inches long (in a large mollusc) with an opaque glassy appearance. It is part of the digestive system of the mollusc, providing certain enzymes such as cellulases for the breakdown of algae. Some styles are full of the distinctive spirochete *Cristispira*. The gel-like viscosity of the crystalline style is the sort of environment in which spirochetes do well, corkscrewing themselves through material that other bacteria would be unable to move through. The crystalline style, especially if it is opaque or cloudy, contains a pure or nearly pure concentration of *Cristispira*. Other molluscs (as well as some sea stars, starfishes, and tunicates) also harbor spirochetes, but not in as distinctive a location as the crystalline style.

One possible outcome of your search for a crystalline style is to find none at all. The style is susceptible to changes in food intake by the mollusc. At low tide, the styles of many molluscs disperse or are broken down. High tide brings with it a supply of planktonic food and a return to filter feeding by the mollusc. Under these circumstances, the style is re-formed, allowing spirochetes to reaggregate there. Molluscs that have been harvested and left to dry in a fish market for even a short time are likely to have no visible styles. You must collect them fresh or maintain molluscs in well-aerated tanks before use. A mollusc without a style, however, is still an interesting preparation for the microscope. Read the next section about what you might see, even if it is not spirochetes.

VIEWING *CRISTISPIRA* UNDER A MICROSCOPE

Take care to choose the best crystalline style that you can find. You want one that is cloudy and neither too hard nor too soft but of a gelatin-like consistency (according to the description by Sagan and Margulis in *A Garden of Microbial Delights*). *Cristispira* is apparently quite particular about the viscosity of this organ and will vacate one that does not have the correct firmness.

Mince up and crush the style and press part of it between a slide and a coverslip. Alternatively, dissolve the style in a small amount of salt water, just enough to cover it, for about an hour and examine a little of that liquid. In a freshwater environment, collect tiny (pea-size or less) floating molluscs from among any duckweed, and crush the molluscs entirely. This will expose the nearly microscopic styles and other organs, and it will release any spirochetes that are present. Pick out the bits of shell before applying a coverslip.

Note that the digestive system of any mollusc is certain to be full of bacteria, even if a style and spirochetes are not visible. Put a little of the contents of the "stomach" area under the scope, and view the rich community of bacteria of many shapes. There may even be spirochetes present, although not in the abundance expected in a good style.

Once you have a good specimen, look for elongated, undulating spirochetes capable of moving either forward or backward with ease. When they are stationary, spirochetes are poised in a helical shape, ready to move again. One characteristic of spirochetes used in their identification is the number of complete helical turns, which in *Cristispira* is 2 to 10 turns. Not easily observed is the crest, or crista, which gives the genus its name. This is a ridge that runs the length of the cell, a sort of distention of the cell that is packed with 100 or more internal flagella (not visible with a light microscope). Most spirochetes have 1 to 3 internal flagella.

It can be a challenge to visualize exactly how the flagella of spirochetes cause the helical motion of the cell. In most motile (non-spirochete) bacteria, the flagella are free appendages that extend into whatever medium the bacteria are in. A tiny, motor-like structure at the base spins the flagellum as though it were a little whip (thus its Latin name). Imagine enclosing this apparatus within the outer membrane of the cell, with one or more flagella at either end, and you have the arrangement found in spirochetes—except that somehow the rotation of the tiny motor translates to a rotation of the entire cell, while the flagella remain more or less stationary.

Using another analogy, imagine a tail wagging a dog, then imagine an *internalized* tail wagging a dog. That is one hypothesis (of microbiologist Peter Greenberg) for how spirochete motility occurs. In fact, the exact mechanism of spirochete motility remains somewhat mysterious, and for spirochetes like *Cristispira*, with 100 enclosed flagella, it is quite difficult to visualize how the coordination occurs. Is there significance to having the flagella attached at opposite ends? Yes! That is how spirochetes move easily in either direction. The flagella may alternate in propelling the cell in one direction or another; alternatively, the flagella may move simultaneously, conferring other types of motility such as flexing in place.

CULTURING *CRISTISPIRA*

Microbiologists have not yet succeeded in culturing *Cristispira*. It appears to be quite fastidious in its growth requirements.

SPIROCHETES OF TERMITE INTESTINES

The anaerobic, cellulose-rich intestines of many termites are wonderful environments for a host of distinctive bacteria. If you are willing to prepare a specimen, this is one of the most reliable ways of observing a variety of spirochetes in action. However, a microscope is necessary. The termite itself harbors hundreds of species of bacteria and cannot be considered a field mark for any particular bacterial group.

VIEWING SPIROCHETES OF TERMITES UNDER A MICROSCOPE

Find a termite, preferably a subterranean type (e.g., *Reticulitermes* in the eastern United States) or a damp wood type (e.g., *Zootermopsis* in the West) or any of a number of dry wood Kalotermitidae (fig. 16.1). If you are an educator, you may have access to termites via biological supply houses such as Ward's, Carolina, and Connecticut Valley. Select a worker with a brownish abdomen, indicating that it is an active eater of wood. Place a tiny drop of 0.6% sodium chloride (NaCl; table salt is okay—try about $^1/_4$t in 1 cup of water) along with a termite on a slide, and lop off the head of your termite with a pair

FIGURE 16.1. Termite hindguts are a wonderful source of spirochetes and many other bacteria, and often contain a host of protists as well.

of forceps or a small blade. Seize the anterior end of the headless termite with the forceps, and press the tip of the posterior end with the edge of a fine needle or pin. Pull the pin gently away from the termite, drawing out the digestive system with it. Discard the termite body, mince up the digestive system, and place a coverslip on top.

Frustrated? Or reluctant? A simpler procedure is to cut off the entire abdominal segment and mince it all up. There will be more debris to look through, but you will still be able to make observations. Be prepared for a surprise.

Termite intestines are absolutely teeming with microorganisms of all shapes and sizes. The largest cells are protists (eukaryotes), which are fascinating in their own right. They are obligate symbionts of termites and may be similar to some of the first eukaryotic cells. Smaller cells (coccoids, rods of all lengths, spirals, and long helices) are bacteria. Any long, straight, spore-containing bacteria that might be present are the gram-positive *Arthromitus*. Look for spirochetes corkscrewing along or staying still for a few seconds,

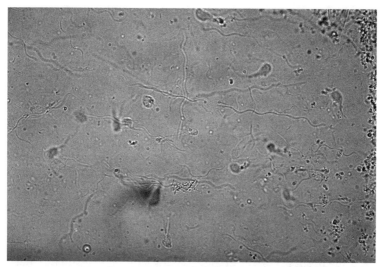

FIGURE 16.2. Spirochetes can be seen in this view of termite hindgut fluid; look for long, coil-shaped bacteria. A few protists are also present.

sometimes flexing their coils a bit (fig. 16.2). Rarely, you may observe a "bloom" of spirochetes: a spectacular sight of hundreds to thousands of them attached to a bit of debris and moving in synchrony. The synchrony does not indicate any deliberate coordination among the spirochetes but rather is a simple way for tightly packed organisms to move. Think of how easy it is to walk along with a crowd but how difficult to turn around and try to go back against the crowd.

In some cases this tendency of spirochetes to coordinate is associated with motility symbioses. These involve protists, especially devescovinids found in the intestines of termites from family Kalotermitidae. Devescovinids are often covered with bacteria, and larger ones are more likely to have spirochetes aboard. Depending on the number of spirochetes and their degree of coordination, some may be capable of propelling their host along. The biologist Lynn Margulis has hypothesized that eukaryotic motility organelles such as cilia may have evolved from symbiotic spirochetes that became permanently attached to their host.

CULTURING SPIROCHETES OF TERMITES

Termite spirochetes have proven to be very difficult to culture; therefore, this is not a project for amateurs. However, you can use your termites as temporary culture chambers to encourage a bloom of spirochetes. Collect some termite workers in which you have observed spirochetes already. Place them in a small jar on a damp paper towel, which (as a wood product) is a food source for them. Keep the jar warm (at about body temperature) for 5 to 15 hours. If you have access to a microbiology lab, simply set the incubator at 37°C. At home, you can improvise, perhaps with a yogurt maker or a warm radiator. Even better is to follow this heat treatment with 3 to 5 (or even more) days of starvation at room temperature. Or you can dispense with the heat treatment and simply use the starvation treatment. How do you starve termites? Remove the paper towel but continue to supply water to them, taking care that you do not inadvertently supply food, such as a cellulose sponge. Supply daily water by placing in the jar a moist sponge made of plastic (presumed to be indigestible to termites).

At the end of whatever treatment you choose, open a termite and make a slide preparation from a small sample. You may see wonderful blooms of spirochetes and also possibly long, spore-bearing rods—the gram-positive bacterium *Arthromitus*. The spirochetes, if they are attached in large numbers to a bit of debris, may be undulating in synchrony. You might also notice that the treatment has removed nearly all the other organisms that you saw in untreated termites. Somehow, the spirochetes must be thriving temporarily on the debris of the dead microorganisms.

Another culturing method exists for the ambitious home-microbiologist interested in experimenting with spirochetes. Prepare for some disappointment, as it is not at all guaranteed to yield good results. Symbiotic spirochetes such as those of molluscs or termites are too fastidious in their requirements for this method. Instead, use freshwater or marine sediments or muds, since few bacteria other than spirochetes can travel freely through these extremely viscous substrates.

Make an agar medium consisting of 1% agar in sterile water or boiled seawater or pond water, and add a very low concentration

(<1%) of nutrients (e.g., sugar or boiled beef broth). (See appendix A for more details on culture methods.) Boil the medium to dissolve the agar and nutrients, then dispense it into small jars or vials (as sterile as possible—try boiling the jars). When the medium is cooled and hardened, try your inoculations. Select a sample (a bit of anaerobic mud, for example) and introduce a few tiny particles well beneath the surface of the agar, using a toothpick to poke them in.

Watch over the course of a week to see if a "veil" of spirochetes has emerged from the particles and has begun to swim out through the agar. A hand lens will help you see what is going on around the particles. You can sample from this veil by digging out a tiny sample with a toothpick. View it under the microscope, and you may be rewarded with the sight of spirochetes.

SUMMARY: FIELD MARKS AND HABITATS OF SPIROCHETES

- digestive system (crystalline style) of freshly collected mussels, oysters, and other molluscs (*Cristispira*)
- intestines of termites

Chapter 17

Thermus and *Deinococcus*

Thermus and *Deinococcus* are tough, resilient bacteria that form their own cohesive taxonomic group based more on their DNA sequences than on physical characteristics. *Thermus* has visible field marks in certain hot springs, whereas *Deinococcus* is completely lacking in field marks.

Thermus

Thermus is a hot spring organism that was isolated originally in Yellowstone National Park from neutral to alkaline springs ranging from 50 to 80°C (122 to 176°F). It is also found in marine thermal springs and even in hot water systems and thermally polluted waters. When isolated and cultured, it may make yellow-pink-red colonies consisting of individual rods or filaments. The colored pigments are carotenoids and are usually present in *Thermus* from habitats that are exposed to sunlight. *Thermus* colonies from dark water pipes are often colorless.

Thermus has a lucrative and practical application as a source for a heat-resistant version of the enzyme DNA polymerase. All organisms use DNA polymerase to replicate their DNA during reproduction. The DNA polymerase of *Thermus*, however, is especially heat resistant, apparently an adaptation to its habitat. It is therefore useful in the polymerase chain reaction (PCR), which is one of the most important biotechnical procedures used today. It is the mechanism by which huge quantities of DNA are generated quite accurately and specifically from a small original sample. The procedure involves cycling the chemical reactants (including DNA polymerase) through a series of temperature changes ranging from about body temperature to near boiling. It takes a tough enzyme to be able to endure that and keep functioning. The polymerase chain reaction is so important—and so profitable—that many patents have been established for the DNA polymerase of *Thermus* and of other thermotolerant organisms. And there is great incentive to find and patent more polymerases of hot spring organisms. Look at certain hot springs, especially in Yellowstone, where researchers are actively pursuing *Thermus* and other species, and you will see what might be considered field marks for this genus.

FIELD MARKS AND HABITATS OF *THERMUS*
Thermus species are found in hot spring environments, with temperatures of 50 to 80°C (122 to 176°F) in neutral to alkaline waters, such as those in Yellowstone National Park. Indeed, Yellowstone is the focus here, as it is in chapters 1 and 3 on thermophilic bacteria, because it is one of the most extensive and accessible hot spring areas in the world.

The original discovery of *Thermus* was made in Octopus Spring near the Great Fountain Geyser in Lower Geyser Basin. If you choose to visit this site, what you will see is an area more of historic importance than of current identifiable field marks. The spring has changed in temperature and conformation since the original discovery. Nevertheless, some people working in the biotech field may wish to make the "pilgrimage" to gaze on the area from which the famous PCR enzyme was isolated (plate 93). Octopus Spring can be accessed by walking up a hill across the street from Great Fountain to a blighted area of unstable, flooded ground. The path is unmarked, and Octopus Spring is not listed in the Park Service brochure. Thus, you should be extra cautious in your trek and should not approach past the edge of the spring.

Elsewhere in the park, you may find areas in which *Thermus* is currently being sampled by researchers. In fact, the Park Service now strictly regulates all the biotech projects and has labeled and described some research areas in brochures. The rangers are also well informed about thermotolerant bacteria such as *Thermus*. Cistern Spring in Norris Basin is one spring now being studied (plate 94). The pool itself is rich in silica and is bright blue in the center, with lovely shades of lilac, pink, and orange around the edge and an overflow of descending terraces of sinter (a silicate mineral). The surrounding area is a "forest" of dead trees, indicating the periodic lethal shifts of hot runoff water. At the lower end of its optimal temperature range, 50 to 75°C (122 to 167°F), *Thermus* is obscured by colorful photosynthesizers. Therefore look for *Thermus* at the high end of the range, 75 to 80°C (167 to 176°F). Cistern Spring itself may be hotter than 80°C, but keep in mind that the runoff is cooler. According to Thomas Brock, in *Life at High Temperatures*, *Thermus* is visible in this pool as hair-like, light brownish-pink filaments. The surroundings may be white and steaming, indicating a temperature high enough to have eliminated most other organisms. You will not be able to get close enough to examine the colored runoff for any filaments, however. Straying off the boardwalk is dangerous in such an area. One possibility is to use field glasses to focus in on a section of the colored sinter to see if pastel-colored filaments against a white background are visible.

DEINOCOCCUS

Deinococcus was originally discovered in canned meat that had been preserved using irradiation (though it is not a pathogen). The most distinctive characteristic of *Deinococcus* is its ability to withstand doses of radiation thousands of times greater than typical lethal doses. Five hundred to a thousand rads of radiation are lethal to humans. A typical dose of radiation used to sterilize surfaces by killing the bacteria is 100,000 to 300,000 rads. Some *Deinococcus* species can tolerate millions of rads.

When cultured, *Deinococcus* forms pink-red colonies; however, it rarely finds enough food in a natural setting to be able to grow abundantly. Rather, it is an extraordinarily resistant opportunist, waiting out dry conditions for years within a tough cell wall—more a part of dust than of the living world. When *Deinococcus* (*deino* means strange or wondrous, and *coccus* means spherical) was first described, it was thought to have a gram-positive cell wall, but the wall is in fact an extra thick and tough gram-negative wall—or perhaps a wall that defies simple characterization as positive or negative. It may be that resistance to desiccation is the primary characteristic of *Deinococcus*, while resistance to radiation is a secondary consequence of the cell's ability to remain dormant under the most inclement conditions.

Other organismal strategies for avoiding damage by radiation (from the sun) and other sources include staying under water or sediments and being pigmented. All organisms have repair mechanisms for DNA that has been damaged by radiation. However, *Deinococcus* seems to be remarkably good at repairing DNA no matter how fragmented it has become as a result of irradiation or desiccation.

FIELD MARKS AND HABITATS OF *DEINOCOCCUS*

There are no good field marks for *Deinococcus*, and in fact it is not at all clear to microbiologists what its natural habitat is. It can be isolated in small numbers from almost any environment, but presumably in most cases it exists in dormant form, patiently awaiting sufficient food and moisture to grow.

Nevertheless, at least one general principle of bacterial field identification may be gleaned from this strange organism. Heterotrophic opportunists usually do not produce good field marks, especially when that heterotrophy involves consuming rich but transient foods that are also of interest to a myriad of other opportunistic heterotrophs—not just bacteria, but also fungi, protists, and even animals. Ironically, these types of heterotrophs, which remain "invisible" in their natural habitat, grow remarkably well on nutrient agar in the lab and are therefore among the best described and studied of bacterial species.

Imagine tiny *Deinococcus* emerging from dormancy just long enough to enjoy a fraction of a crumb of moist food, then quickly returning to deep, safe inactivity. *Deinococcus* is everywhere, but that means it is nowhere in particular and therefore fails to produce the usual colorful, smelly field marks featured in most chapters of this book.

SUMMARY: FIELD MARKS AND HABITATS OF *THERMUS*

- hot springs or runoff from hot springs with temperatures of 50–80°C (122–176°F) and neutral to alkaline conditions
- yellow-pink-red filaments (may be obscured in lower temperature ranges by colorful photosynthesizers; see chap. 1)
- deep-sea thermal springs
- hot-water systems (difficult to detect because pigmentation is usually lacking)

Note: *Deinococcus* does not have any good field marks.

Chapter 18

Planctomycetes and Their Relatives

The most distinctive members of the planctomycetes are the stalked bacteria, which have stemlike protuberances, or stalks, with which they attach to substrates or to each other. Planctomycetes is a diverse group, and little by little DNA sequence analysis has revealed who is related to whom—with some surprises. For example, *Chlamydia*, a stalkless genus that is an obligate inhabitant of other organisms' cells (usually with pathological consequences), turns out to be a relative of the planctomycetes. Because it is always within other cells, *Chlamydia* is beyond the purview of this book, but it provides a good example of just how diverse a group of bacteria can be. *Chlamydia* and stalked bacteria bear no morphological or functional resemblance to each other. Yet it is true of many bacterial groups (such as the gram-positives and the proteobacteria) that if there are free-living, opportunistic heterotrophs, there are also members of the group that have taken opportunism to an extreme and evolved as pathogens.

Another surprise resulting from an analysis of DNA sequences is that having an appendage is not a distinctive enough characteristic to place bacteria securely within the planctomycetes. Some beta proteobacteria (e.g., *Gallionella*) have stalks (see chapter 7). Some alpha proteobacteria (e.g., *Caulobacter*, *Hyphomicrobium*, and *Pedomicrobium*) as well as some gamma proteobacteria have appendages of a different kind called prosthecae. (Stalks are extensions of a cell that contain no cytoplasm, whereas prosthecae are extensions that do contain cytoplasm.) Prosthecae are not found in the planctomycetes. Stalks such as those of some planctomycetes are a mechanism for maintaining position in an aquatic environment. The bacteria can attach themselves to rocks or other substrates in the water and by forgoing motility can take advantage of the motility of their environment.

Attachment of one sort or another is a common adaptation in many groups of aquatic microbes. Although bacteria are often motile via flagella, this type of locomotion is not a very effective way of maintaining position in water. Any turbulence or current is bound to be a far greater force than that produced by flagella, and because of the small size of bacteria, the viscosity of water can be overwhelming. An analogy is that a bacterium swimming (or struggling) through water is equivalent to us swimming (or struggling) through a pool of soft gelatin. Most bacteria find themselves making little forward progress but instead slowly sinking through the water column until they reach bottom. Attachment is often more efficient than motility for maintaining position.

Stalks are also sometimes used as a flotation mechanism: a group of stalked bacteria can form a sort of rosette of stalks attached to each other that has enough surface area to remain suspended in a water column. The design may be thought of as something like a sea anchor. Gas vacuoles or vesicles are used for flotation by planktonic members of several groups of bacteria, but not the planctomycetes.

FIELD MARKS AND HABITATS OF PLANCTOMYCETES

Planctomycetes, along with other stalked organisms, can be isolated from many soils and all sorts of bodies of water—lakes, ponds, rivers, and streams as well as saline or marine waters and hot springs. However, because such environments do not usually have high concentrations of nutrients, it is difficult for planctomycetes (which are heterotrophic) to build up large enough numbers to produce field marks.

When a body of water is eutrophic or part of a system that receives sewage or other nutrient-rich runoff, then planctomycetes might well bloom on surfaces and through the water column. Certainly they can be isolated from such waters. However, many other groups of heterotrophs and autotrophs also proliferate under such circumstances, obscuring each other's field marks. Therefore, plan to infer the presence of planctomycetes based on the following description of surface films, and if you have a microscope, try trapping some, as described below. (Note that nutrient-poor, iron-rich environments with films of manganese and iron oxides may be good habitats for the stalked beta proteobacterium *Gallionella*; see chapter 7.)

Although rocks are an obvious substrate for stalked bacteria, there is another, more intriguing, one, easily found on the surfaces of still waters. Look for a surface film, sometimes a little dusty or crusty in appearance although extremely thin. The film may be cohesive enough to be folded and wrinkled by wind and waves. It may also appear bubbly or even foamy if it is rich with organic materials. Such films are a wonderful microbial substrate, and many bacteria and protists are likely to be suspended beneath it, dangling into the water. The community of organisms in the surface film (whether or not you can see it) is called a neuston (plates 95–97).

VIEWING PLANCTOMYCETES UNDER A MICROSCOPE

Suspend some clear glass microscope slides in a body of water of your choice (either still or running) and wait from several days to 1 or 2 weeks (plate 98). Pull the slides up and wipe one side clean while keeping the other side quite wet. Add a coverslip to the wet side, and take a look.

You can also try to collect a bit of the surface film from a still body of water. Try bringing a slide up underneath the film to capture a bit, or placing a slide flat down on the film to lift some of it off. Dry the bottom of the slide, and place a coverslip on the moist side.

In either preparation, you might well see many large, stalked organisms bouncing up and down quite energetically on their spring-like appendages. Don't get too excited. These are not bacteria but rather stalked ciliates (eukaryotes), although they deserve to be studied in their own right. You may also see rotifers: attached, telescope-like, multicellular animals with two rotating wheels of cilia. These are also worthy of study. Furthermore, the slide will probably be full of eukaryotic algae, including diatoms and many nonstalked ciliates and other protists. Remember that bacteria are usually one to two orders of magnitude smaller than a eukaryotic cell.

After lingering in this world of stalked ciliates and rotifers for a while, pull yourself away and descend to the level of bacteria

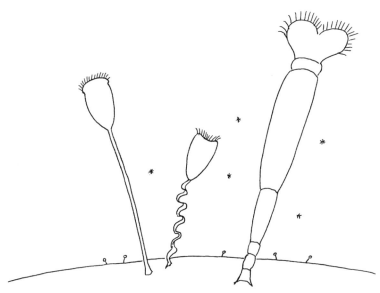

Figure 18.1. The attached neuston community may include many large organisms, including rotifers and stalked ciliates. Look for much smaller stalked bacteria in the background.

(fig. 18.1). You may need to dim the light a bit. Many bacteria should have attached both to your slide and to the surfaces of the eukaryotes themselves. Attached bacteria might include photosynthesizers if you left it in a photic zone. You may see stalked bacteria, some of which could be planctomycetes, although it will be difficult to distinguish them from bacteria with prosthecae. Stalks can be extremely fine to the point of being invisible even under high power. Slightly easier to observe are the rosette-like arrays of stalked bacteria attached to each other, most likely planctomycetes.

SUMMARY: FIELD MARKS AND HABITATS OF PLANCTOMYCETES

- filmy, sometimes dusty surface on a body of still water, poor to moderate in nutrients (usually a collection of bacteria, making unequivocal identification difficult)

Appendix A

How to Culture Bacteria

Do you, as an amateur naturalist, need to culture bacteria? The positioning of these instructions in the appendix is meant to give a clear message: Culturing bacteria is an optional, extra step. You do not have to culture bacteria to appreciate them. If the process sounds like something you would rather not do in your kitchen, then don't do it! For those who decide to try some of these techniques, please read appendix B on safety. These projects are appropriate for enthusiastic but cautious amateurs.

A Little History

There are two major methods for culturing bacteria, each associated with a different tradition and philosophy:

1. *Pure culture methods*, in which the goal is to grow a single species of bacteria with no "contaminants" (other species) present
2. *Mixed culture methods*, in which assemblages of bacteria are provided with conditions as close to "natural" as possible so that they may grow and interact together

Pure culture has been predominant in Western microbiology and has been essential in the identification and characterization of bacterial species. Mixed culture methods were pioneered by Russians and until recently predominated in Russian microbiology; it is by means of these mixed culture methods that microbial ecological relationships have been deciphered and analyzed.

The pure culture methods outlined in this appendix are similar to the ones worked out in the late nineteenth century by the German biologist Robert Koch, presumably with assistance from his wife, Emmy, as these procedures initially took place in their kitchen. Koch was eager to isolate single bacterial types so that he could determine the causes of certain diseases. He discovered that individual, distinctive colonies of bacteria would grow on the cut surface of a boiled potato. Other solid surfaces also gave good results, most notably, an aspic of gelatinized beef broth. Credit for this discovery is attributed to Lina Hesse, the wife of a coworker of Koch. Surely many other people (scientists and nonscientists) had noticed some sort of growth on aspics and other moist surfaces of food that had sat around uneaten for too many days. However, Koch (prompted by Hesse) was the first to recognize the practical significance of being able to grow individual colonies on hard nutrient surfaces. Modern Western microbiology was initiated using this technique and has been conducted in this manner ever since.

The mixed culture method was developed in the 1880s by the Russian microbiologist Sergei Winogradsky, who cultivated samples from the environment under controlled conditions in order to understand "the big picture." How are bacteria interacting with each other? How are they changing and using the chemicals in their environment? Thus began environmental and soil microbiology. The most famous method for mixed culture is called a Winogradsky column; instructions for making one are given at the end of this appendix.

Other methods of controlled mixed culture come out of culinary traditions such as sourdough baking, wine and beer making, cheese making, and pickling. Traditional agriculture also involves the (often naive) use of microbial communities in techniques such as composting and rotating fields with leguminous crops as well as maintaining healthy ruminant animals such as cows. The chapters on gram-positives provide several suggestions for making cheese, beer, compost, and so forth, though you should rely on other books for exact recipes and procedures.

Pure Culture Method and Sterile Technique

First, a disclaimer: these instructions will enable you to make as pure a culture as can readily be done in your kitchen. You most likely do not have access to a high-pressure steam sterilizer (an autoclave) and therefore cannot keep your glassware and culture medium absolutely free of contaminating bacteria. These instructions call for boiling your medium and glassware, but you could use a pressure cooker; follow the instructions for the particular brand you have. Gram-positive bacteria and a few others can survive boiling temperatures and may turn up as contaminants; consider them to be interesting additions to your experiments. Indeed, you may consider survival of boiling a reasonably reliable indicator of the gram-positive nature of a bacterium.

Furthermore, you will not be using truly sterile (autoclaved or flamed) instruments but will take advantage of something well known to lab and field microbiologists. Packages of cotton swabs and toothpicks that have not been overly exposed to the air are, for many purposes, "sterile." Keep in mind, also, that bleach kills any microbe, so dousing with bleach will be your method of disposing of any cultures you do not want.

ABBREVIATED PROCEDURE
Pour into "sterilized" (boiled) jars (e.g., baby food jars) $^1/_4$ to $^1/_2$ inch of hot "sterile" (boiled) dilute nutrient broth to which a solidifier has been added. Koch used gelatin; agar is used today in most microbiology labs and gives better results. Once the broth has cooled and is solid, you can use it to isolate and visualize bacterial colonies. Just swab on a sample using a cotton applicator or toothpick, put a boiled lid on, and wait a few days to see what grows.

DETAILED PROCEDURE
1. Boil 6 to 8 baby food jars and lids in a large saucepan or a pressure cooker and let them drain—while still hot—upside down on paper towels. Boil a 1 cup measure and a pint jar for mixing.
2. Decide on a source of nutrients. There are many choices—this is the creative step! In part, your decision should be based on an attempt to "think like a bacterium." Bacteria that are normally active in a particular environment tend to be rather specialized in their use of nutrients and other materials. Growing a specialized bacterium can sometimes tell you something about the microbial activity going on in the environment from

which you isolated it. Growing an opportunist or generalist is a little easier but is less creative and gives you less information. Opportunists respond well to rich, abundant food. Most likely they were resting in a dormant state until you provided them the food. Therefore, consider using a relatively dilute, specific medium to tease out the active microbial members of a community. Don't provide too much of a banquet.

Nutrients might include small quantities of soil or compost, dry or fresh vegetation, roots, feathers, manure (especially from herbivores), and even minerals. In all cases try amounts ranging from fractions of a teaspoon to a couple of tablespoons per 2 cups of water. You might also try to make a culture medium using water alone. There may be enough trace nutrients to sustain some interesting bacteria. Indeed, for photosynthesizers and chemosynthesizers, you want no nutrients at all; let them make their own food.

3. Choose your water. It could be tap water, pond or river water, or even distilled water. For marine bacteria, it could be ocean water or Instant Ocean from an aquarium supply store. If you use Instant Ocean, you can experiment with different salt concentrations.

4. Combine 2 cups of water plus a small amount of your chosen nutrient (if any) in a saucepan (or pressure cooker) and boil for a few minutes. How long? Some nutrients are destroyed by lengthy boiling, while others are released and become more accessible only after boiling. Simply experiment, perhaps starting with 3 to 5 minutes. If any particles remain in the broth after boiling, filter the broth through a coffee filter and boil a little more. Alternatively, don't boil but steep the nutrient in freshly boiled water (covered) for minutes to hours.

5. Choose either agar ("seaweed gel") from the health food store or gelatin from the grocery store to solidify the broth. Agar is preferable, although it may be more expensive and difficult to find. Other solid media include pectin or potatoes.

If You Use Agar: Measure out 1 cup of broth into a saucepan and sprinkle on about 1 teaspoon of powdered or about 1 tablespoon of flaked agar. Let the mixture soak for a few minutes, covered, then boil for about 4 to 5 minutes. Pour the hot mixture about $^1/_4$ to $^1/_2$ inch deep into the boiled baby food jars and cover with the lids. Refrigerate until use.

If You Use Gelatin: Cool $^1/_2$ cup of broth in your boiled pint mixing jar. Sprinkle one packet (1 tablespoon) of gelatin on the broth and let it soften and partly dissolve for a few minutes, covered. Meanwhile bring the rest of the broth back to a boil. Measure out $^1/_2$ cup of the boiling broth and add it to the gelatin mix to make a total of 1 cup. Swirl the jar to mix and pour the gelatin into boiled baby food jars $^1/_4$ to $^1/_2$ inch deep. Cover and refrigerate until use.

Other Solid Media: How about pectin? Pectin, the other major culinary solidifier besides gelatin, which is used to make jams and jellies, may seem like a logical choice for bacterial culture. However, pectin requires a lot of sugar to work—so much that it tends to inhibit bacterial growth. That's why jams and jellies tend to keep pretty well. However, if you happened to be making jelly anyway, you could pour some pectin into a small jar and experiment with it, to see what bacteria would grow. Fungi (molds) seem to be more tolerant of high sugar, so you might end up culturing those.

How about potatoes? Good idea and very simple. A cooked or freshly cut raw potato is a ready-made substrate for bacterial growth. That's what Koch first used. Just drop it into your "sterile" jar and moisten it with a little boiled water or boiled nutrient broth.

6. Inoculate your solidified medium. Choose an inoculum, such as a few drops of pond water or a swab from your armpit or the surface of a leaf. If the surface is dry, first moisten the cotton swab with a little boiled water. Spread the invisible microbes lightly on the surface of the medium. Try not to break the surface; if you do, however, you will be able to find out whether your microbes can live in the relative lack of oxygen under-neath the surface, or whether some microbes can move around in the dense substrate. A toothpick stabbed under the surface is a good way to deliver microbes to those depths. H. Steven Dashefsky, author of several excellent books on science fairs, suggests having an insect walk across the surface to see what was on its feet. When you are done replace the lid either

loosely or tightly. A loose lid lets in more oxygen and favors different species.

7. Choose incubation times and conditions: sunlight? darkness? warmth, such as near a radiator? room temperature? Think about what variables you might control, and experiment with these. Incubation time can range from one to several days. If you are using a gelatin-based medium, the incubation time should be shorter because many of the bacteria you are likely to culture can dissolve gelatin, turning it into a soupy mess. You need to look at your results before this happens.

8. Examine what grew: after a few days, look for colonies on the medium. Shiny, droplet-like colonies of various sizes, shapes, and colors are probably bacteria. However, some yeasts make similar sorts of colonies. Fuzzy colonies are probably fungi, although small ones could be gram-positive actinomycetes.

If you have many colonies growing in and on each other, try using a more dilute inoculum the next time, or try a more dilute (less encouraging) medium. You might even try adding to the medium a small quantity of something that could inhibit growth, such as a bit of sawdust from pressure-treated wood or some household disinfectant. Try soaking bits of paper towel with a weak solution of liquid disinfectant and placing those on the surface of the agar. You might be rewarded with the sight of a zone of inhibition around the paper towel, where no microbes can grow. You can also try boiling your inoculum to select for just the spore-forming gram-positives that resist boiling. Or you can observe the crowded colonies and try to figure out whether any of them are either inhibiting their neighbors (perhaps by means of antibiotics) or enhancing the growth of their neighbors (perhaps by providing nutritious wastes).

If you have little or no growth, try increasing the concentration of nutrients, or try a different inoculum. However, sparse growth of one particular type of colony can be seen as a successful culture method for a specific microbe.

9. If you have a microscope, use a toothpick to pick up a dab of material from a colony. Place it on a slide in a small drop of water, and add a coverslip. Look at the slide under high power

(e.g., 400×), adjusting the light. You may well see rods or coccoids, two of the most common bacterial shapes. If the cells are large and easy to see, and especially if they seem to be made up of long filaments, you are probably looking at fungi.

MIXED CULTURE METHOD
THE WINOGRADSKY COLUMN

The most famous mixed culture method is the Winogradsky column. Making a Winogradsky column is an excellent project for home or classroom. With little maintenance, a well-established column can last for years. This is a safe experiment; unless you add a large quantity of rotting meat, you are probably not cultivating pathogens. In a sense, you are setting up a specialized compost heap, which can also be done on a small, selective scale on your windowsill (see chapter 11).

In brief, a large jar is prepared with some crumpled paper towel (as a source of cellulose) in the bottom and topped off with rich black sediment from an estuary or mudflat along with some ambient water. After a few weeks to months of incubation, partly or entirely under a light source, a colorful community of sulfur-loving bacteria should develop. The culturing technique is so integral to understanding sulfur bacteria that the detailed description is provided in chapter 9; variations on the technique for the encouragement of metal-oxidizing bacteria and for cellulose-consuming gliders are described in chapters 7 and 15, respectively.

A basic recipe for a Winogradsky column (although Winogradsky himself would probably not have stated it quite this way) is as follows:

1. Find a large jar, preferably tall and cylindrical.
2. Add moderate amounts of ingredients that you hypothesize could be of "interest" to some bacteria. These might include
    straw
    dry leaves
    bits of bark
    pieces of metal
    horse manure
    bits of insect or lobster shells

hard-boiled egg yolk

salt

Epson salts (a source of sulfate)

something acidic or alkaline

You can try just about anything. Use your imagination—think like a bacterium!

3. Top off with mud or soil along with ambient (or other) water.
4. Decide on incubation conditions:

covered or dark (or create some of each by masking part of the jar with black paper)

warm or cool

tightened lid or lidless

compensation for evaporation (by adding water) or not

other parameters—use your imagination

5. Watch for layers of colored bacteria to form over a period of days to weeks. Take samples for microscopy from specific layers, taking care to remove any bits of sand that might damage your coverslip. Plan to keep the column for months or even years. The author has had a Winogradsky column decorating her windowsill for more than twenty years.

Percent Solutions

The instructions above tell you how to make solutions using teaspoons and cup measures. In some cases, perhaps for a science fair project, you might want to be able to describe a solution in more scientific terms, as an "X% solution," or you might be using instructions from another source that tell you to make a certain percent solution. Here are some guidelines for using kitchen utensils to create specific types of solutions.

If your kitchen measuring utensils and scales are calibrated in metric units, then the method for making solutions of a certain percentage weight per volume or volume per volume is easy. For example, 1g sugar in 100ml of water is a 1% solution (weight/volume). One ml vinegar in 99ml water is a 1% solution (volume/volume). Even if you have just a set of measuring spoons and cups calibrated in English system units, you can make solutions with certain percentages of ingredients. Keep in mind the following approximate equivalents:

$3t = 1T$

$2T = 1oz = 30g$

$8T = \frac{1}{2}c$

$2c = \frac{1}{2}L$

Let's say you would like a 1% infusion of soil in water. One cup of water $= 16T = 48t$, so you would add 0.48t soil. Round that off to half a teaspoon, unless you have more precise ways to weigh and measure. Half a teaspoon of an ingredient in 1 cup of liquid, then, yields an approximately 1% solution.

Safety Precautions

The caution that one should not drink or eat one's bacterial cultures normally would go without saying. However, several sections of this book elaborate on the roles of bacteria in enhancing the flavors and aromas of food and drink. Therefore, you may be tempted to grow bacteria for the purpose of experimenting with cuisine. Please do not do so unless you are following clear recipes from manuals on cheese making, wine making, pickle making, or the like. This field guide is not intended to provide instructions on how to prepare microbially enhanced foods and drinks.

The tone of this field guide may suggest to some readers that nearly all bacteria are harmless. They are not. Many—especially proteobacteria and gram-positives—are opportunists that can become pathogenic if they gain access to the nutrient-rich conditions of your body. If you choose to culture bacteria, take care not to breathe them in or get them on your skin. Use culture methods as described in appendix A, which are meant to discourage opportunists by using relatively dilute nutrients. This approach better approximates most natural conditions and gives you a better idea of the activities of indigenous bacteria. Using dilute nutrients discourages opportunists, although it does not eliminate them completely. An example of an opportunist is *Staphylococcus*. This species is indigenous to our body surfaces and cavities; however, a *Staphylococcus* infection in which the bacteria have gained access to internal organs, especially if the immune system is not functioning properly, can be fatal.

Take these precautions:

1. Do not eat or drink near your bacterial cultures. This means that while you can prepare bacterial culture medium in your

kitchen and store unused medium in your refrigerator, you should keep your living cultures elsewhere.

2. Keep the lids on your cultures, or remove the lids only briefly as needed. Do not take a deep breath close to uncovered cultures.

3. Wash your hands well after working with bacterial cultures.

4. Wash with rubbing alcohol whatever work surface you are using before and after working with bacteria.

5. When you are done with your cultures, pour in a little bleach. That should kill everything so the cultures can be safely tossed out.

6. While it is common practice in high school and college labs to use rich nutrient agar to culture bacteria, use more dilute nutrients at home. The main reason that nutrient agar is so commonly used in class and science fair projects is that it yields quick and easy results: lots of hardy opportunists. Much more interesting, however, are the slower growing specialists that are normally active in a particular environment. Such bacteria are much less likely to have pathogenic capabilities.

7. Use a "Russian style" mixed culture method (described in appendix A), which does not provide as favorable conditions for pathogenic opportunists.

8. If you have a compromised immune system, your doctor has probably instructed you about what precautions to take. You should not culture any bacteria at all without the permission of your doctor.

9. If you are doing a classroom or science fair project, you are required to follow the guidelines issued by Science Service, Inc., 1719 N. Street N.W., Washington, D.C., 20036; phone: 202-785-2255. If you are working at home on a science fair or classroom project, those same guidelines are applicable. It is suggested that you request a complete set of the guidelines.

How to Use a Microscope

I beheld with wonder many little animals of divers kinds, which escape our naked eye.

ANTON VAN LEEUWENHOEK, 17th-century
Dutch microscopist, as quoted by C. Dobell

To view most bacteria, you need a compound microscope with at least one high-powered objective lens in the range of 40× (40 times magnification). The label "40×" is usually engraved on the side of the lens. If the eyepiece of the microscope is labeled 10× (10 times magnification), then the combined power of eye piece and objective is 40 × 10 or 400×. Even better is a microscope that also has a 100× objective, giving a combined power of 1,000×. Such microscopes are not likely to be found in toy stores, popular science stores, or museum shops, however. The objectives on less expensive scopes typically give you combined magnifications of 200× to 250× or less, sometimes much less.

Where can you find a microscope with a magnification of 400× to 1,000×? You may be able to get access to one from a local high school or college, or sometimes schools and universities that are buying new microscopes are willing to sell their old ones. Used microscopes are also sold at flea markets or through classified ads. It could be worth buying an older scope with the right objective lenses if you want to do a lot of microscopy at home.

Lighting is another factor to consider in choosing a microscope. If you find an old, sturdy scope with no plug-in light but just a little mirror positioned to catch the light, consider it, especially if it has the right lenses. Otherwise, there should be an electric light source. Check to see that it works, keeping in mind that frayed cords and loose connections can be fixed easily. If you are cleaning up an old scope, use only lens paper to clean the lenses.

TESTING THE PARTS

So now you have a microscope. Or you have borrowed one. Or perhaps you are at a flea market and have somehow found an electric outlet and are trying out a microscope you might buy. Take a close look and try to identify and test the various moving parts.

FOCUSING KNOBS
The microscope should have 1 to 3 pairs of knobs. Check to see how they work. One pair of knobs (perhaps the only pair or the most conspicuous pair) raises and lowers the stage. These are the focus knobs. There may be a smaller set of knobs inside the focus knobs, that do the same thing but in much smaller increments. These are fine focus knobs. As you try the focus knobs, take care that you do not cause the stage to collide into an objective lens.

If there is a second pair of knobs, they are likely to be used for focusing the light. You should find that they move a small lens (condenser lens) system beneath the stage. In general, you should adjust the condenser lens so that it is in the highest position.

MICROSCOPE WITH A MECHANICAL STAGE
If there is a third set of knobs, both situated to one side of the scope, they may move the stage from side to side and forward and back. This arrangement means you have a mechanical stage, which is a nice option on a microscope, enabling you to scan a slide easily.

PHASE-CONTRAST MICROSCOPE
If you have a moveable condenser lens beneath the stage and it seems a good deal more complex than the one described here—specifically, if it seems to consist of a turntable of different lenses—you may have a phase-contrast scope. The word "phase" may appear on some of the objective lenses. These are useful scopes to have in a microbiology lab; however, they are complex enough that a general set of instructions will not adequately explain how to get the lighting right to view bacteria. If you are using a phase-contrast scope at a high school or college, get help from a teacher or professor on the correct use of the condenser. If you have a phase-contrast scope and

no one to assist you, turn the condenser wheel until it is in a position in which you have an unobstructed field of vision. Use it at that setting until you can read the manual or get assistance.

VIEWING A NEWSPRINT "SPECIMEN"

Now you have played with the knobs. Next, prepare a specimen for viewing. Do not commence by looking at bacteria if you are a beginner. Newsprint is a good subject for your first observations. Have available a bit of newspaper with some printed letters, a glass slide, at least one thin glass coverslip, and some water. Make a "wet mount" by placing a tiny scrap of newspaper on a glass slide, adding a small drop of water, and placing a coverslip on top. Try to always prepare your slide using a coverslip. Flat preparations, as these are called, are easier to view and pose less risk of getting an objective lens wet. You can buy slides and coverslips at a popular science store, or you may be able to get a few from a biology teacher. If you are careful, you can use them over and over.

Clamp the prepared slide onto the stage using the clips or the slightly more complicated mechanical stage setup, if you have one. Always start with the lowest power (the shortest objective lens), even when your goal is to observe tiny bacteria. Look through the eyepiece and begin to use the focus knob. Newspaper print is so large and distinctive that you should be able to get some letters into focus right away. Note that they are upside down and backward. Gently move the slide from side to side and back and forth to get a feel for the effects of the reversed image. The lowercase letter *e* is an especially useful example.

ADJUSTING THE LIGHT LEVEL

It is important that you become comfortable with light adjustments, as these are critical for observing bacteria. How do you adjust the light? This depends on how complicated your scope is. Here are some of the possibilities:

1. If you have a mirror to reflect light from some other source, try tilting the mirror and adjusting the distance and angle of the light source, which might be an ordinary table lamp. In some configurations, the image will be dark; in others, the light

will glare too brilliantly. Find an optimal middle point. One advantage of this setup is that you can use your scope outdoors in sunlight.

2. If you have a built-in light bulb with an on/off or dimmer switch, use that to adjust the amount of light reaching the stage.

3. Whether you are using a light bulb or a mirror, there may be a lever that opens and closes a diaphragm beneath the stage. Use it to optimize the light level.

If all looks fine at the lowest power, rotate the next highest power lens (the next longest) into place. You should find that the newsprint is almost in focus with this lens and that you need to do only a little adjusting, usually just with the fine focus. However, you may need to adjust the light. Remember that adjusting the light does not usually entail moving the condenser up and down. Keep that in the top position. Use all of your objective lenses in turn up to the 40×. Pause there before using the 100× (if you have one).

USING THE OIL (100×) OBJECTIVE

The 100× objective is the most powerful (and most expensive) lens on any microscope. It can be used only with immersion oil—never without. The word "oil" may appear printed on the side of the lens. Using the lens without oil could scratch it. Immersion oil has been formulated to be free of solvents, acids, and other contaminants that might harm your scope. You should be able to obtain immersion oil from a high school or college. If not, resist using your 100× lens until you are able to obtain the oil from another source.

Here is how you use the oil lens:

1. Prepare a wet mount, taking care that it is not too wet or too thick.

2. Use a toothpick to transfer a tiny drop of oil to the top of the coverslip.

3. Peering at the oil lens from the side, gently lower it (or raise the stage) until the lens just touches the oil, causing the oil to spread a little.

4. Look through the eyepiece and *very gently* focus up and down, until an image is in focus. Remember to adjust the lighting too.

What can go wrong with your oil lens? You can crack it by focusing it too vigorously so that it collides with the slide. You can clean off the oil with a tissue or other rough paper, scratching the lens. Use only lens paper, and consider letting a little oil remain on the lens if you are planning to use it again soon. Try not to get the oil on the other lenses or on the stage.

VIEWING REAL SPECIMENS

After your practice explorations using newsprint, get ready to look at real field samples that include bacteria as well as other micro-organisms. Murky pond water is a classic choice. Collect some in a jar, including a few leaves and some sediment, and set it on a windowsill. Over the course of a few days, a microbial community will develop. Take a small drop plus a little scraping from a leaf surface, and make a wet mount. Start with low power, adjusting the light. If you do not immediately find microbes, focus on some debris.

At the lowest power, whatever organisms you see are likely not to be bacteria but rather some sort of protists (e.g., a ciliate) or small animal (e.g., a rotifer or a nematode). They are fascinating in their own right and deserve hours of observation. However, the goal of this appendix is to get you to see bacteria. Adjust the magnification incrementally up to 40×, stopping at each lens to focus and adjust the light.

At 40× (a combined magnification of 400× with your 10× eye piece), you have the possibility of seeing bacteria. Keep in mind that most bacteria are tiny colorless rods and coccoids (dashes and dots) that will not be moving much on your slide. There are exceptions—large, colorful, active bacteria—but those are found only in specific environments, as mentioned throughout this field guide.

Right now, however, you are probably looking at a sample that is full of rather ordinary dashes and dots, 10 to 100 times smaller than the protists. Dimming the light a bit might make them seem a little less transparent. Also, try allowing the slide to sit for several minutes before you view it. This gives the bacteria a chance to make contact with the slide and makes them easier to see.

Frustrated? Pond water is a great introduction to the microbial world and is a good way to start using your microscope on real field samples. However, the bacteria in your sample may prove disappointing or frustrating, as they sometimes are even to professional microbiologists.

You might want to try another sample. Take some beef or chicken broth, noncreamy style with as little fat as possible. Set it out in an open container on your counter for 24 hours or more. Make a wet mount of a drop of broth, and view it with each of your objectives, lowest to highest. At the low powers, you may see nothing, but focus anyway on a bubble or bit of debris. At 40×, you may find your sample teeming with bacteria, including rather long rods and active motile forms. Enjoy! Your soup sample is thoroughly contaminated with opportunistic bacteria that were dormant in your kitchen until you provided this feast.

TIPS FOR VIEWING FIELD SAMPLES

At this point, you are ready to follow the instructions given throughout this guide for collecting and observing particular samples, in the sections titled "Viewing under a Microscope." One admonition, repeated frequently, is to avoid making preparations that contain any sand whatsoever. You must either pick out the bits of sand, grain by grain, or dilute your sample away from the sand. Otherwise, you will crack your coverslip trying to make a flat mount. Also, keep in mind that less is more with bacterial preps; do not use large, overflowing drops. Always begin your focus sequence at the lowest power, even if it means focusing on debris. If you cannot focus at low power, you will have even greater difficulty at higher powers. Adjust the light, taking into consideration that your subjects are often transparent. Also, most preparations benefit from being allowed to settle for several minutes so that the bacteria can make contact with the glass slide. Having most of your bacteria in the same plane of focus, against the slide, makes them easier to see.

The combination of microscopy with the culture methods described in appendix A and throughout the book can be powerful. A little dab of a cultured bacterial colony or film or scum, taken up on a toothpick and prepared with a small drop of water, will likely show you hordes of tiny cells, perhaps with one particular type predominating.

APPENDIX D

Suggestions for Science Fair and Classroom Projects

Consider a different route from that taken by so many science fair participants: don't test mouthwash or soaps or any of the household or personal cleaning products by culturing mouth and skin bacteria before and after. That is, don't do this unless it truly excites and intrigues you and you are willing to ask additional questions and make it more of an exploration. I write this as a judge of science fairs. I've sometimes found that nearly all of the microbiology projects at a science fair are some variation of a before and after testing of household products.

Instead, be more adventurous. After all, you have *A Field Guide to Bacteria* in your hands, which is in itself an unconventional approach to bacteria. Skim through and choose a bacterial group because it seems interesting. Then read the whole chapter for that group, taking notes as you go. What questions do you have that seem to be unanswered, at least by this book? Try to come up with several. Later, when you are considering experiments, some questions may be answered with further reading, whereas others might be too complex for a short-term project. Settle on what seems to be a good question, such as "What conditions are favored by the big, white, conspicuous bacterium *Beggiatoa*?" You know from reading chapter 9 that it likes a certain amount of hydrogen sulfide and oxygen. How quickly can it move to a new position to be in the right gradient of these chemicals? The question could be approached both by experiments in the field and by designing Winogradsky columns with varying conditions. Where do you find *Beggiatoa*? That's why you have this field guide.

A Life List of Bacteria

When you think of a life list, you probably think of birds. Professional ornithologists and enthusiastic birders keep them and take great pride in new sightings. For other organisms, life lists of a sort can take the form of photographs or sketches (such as for mushrooms), dried pressings (such as herbarium mounts), or collections (e.g., of insects or fossils).

Collecting, photographing (unless you have special lenses), and sketching do not work well for most bacteria; however, a life list is feasible. The idea occurred to me when I read that the Dragonfly Society of the Americas had been successful in popularizing dragonfly watching by coining attractive new common names for many species and by encouraging life lists. Perhaps a life list of bacteria might be appropriate too. Indeed, many professional field microbiologists who are familiar with the field marks in this book probably keep an informal life list in their heads: if you name a bacterial group, such as *Beggiatoa*, they could probably tell you exactly where to go to see a nice bloom of it.

Your list might include "first sighting" or "best sighting" (date and location) of every bacterial group and subgroup in this book. One of the ironies of bacterial field marks is that some of them are so conspicuous that you have seen them all your life, yet your "first sighting" is really a first realization that a field mark is connected to bacteria. For example:

May 1999. Freshly plowed cornfield, Rehoboth, MA—smell of geosmin—actinomycetes

You've smelled freshly turned earth before. This was your first recognition of that smell as a field mark.

Other items on the list might be genuine first sightings. Most naturalists do not walk head down peering deep into murky pools or scrutinizing scummy layers of surface film. This book may introduce you to that pleasure for the first time.

Glossary

16S sequence of DNA gene that codes for part of the ribosome, an essential protein-producing complex found in all cells. "S," which stands for Svedberg, inventor of the ultracentrifuge, is a unit based on size, shape, and density of a particle as determined from its behavior in an ultracentrifuge. 16S sequences are useful for constructing phylogenetic trees because they change relatively little over millions of years.

acetyl coenzyme A a large, complex, energy-rich compound used as a starting point for some important chemical reactions, including those by which bacteria use or, in some cases, make food molecules.

acid a chemical with low pH (usually sour), which yields hydrogen ions (used by some bacteria) when in water. Each acid has a corresponding salt form when it is not dissolved in water, indicated by the suffix "-ate" (e.g., acetic acid becomes acetate). Acids are waste products of some bacteria, such as fermenters, and provide food for others, such as those that use oxygen for respiration. Some bacteria use hydrogen ions as part of their metabolism.

acidophile an organism for which an acidic environment is optimal (acid lover). Extreme acidophiles, such as some archaea, can tolerate a pH less than 3. Acid-tolerant organisms, by contrast, can endure acidic conditions but do not find them optimal.

alga(ae) photosynthetic eukaryotes of the protist kingdom.

alkaliphile an organism for which an alkaline environment greater than pH 8 is optimal (alkali lover). Alkaline-tolerant organisms, by contrast, can endure an alkaline environment but do not find it optimal. Many alkaliphiles are also halophiles.

anaerobic lacking or not requiring oxygen; can describe either environments or organisms. "Anoxic" also refers to a lack of oxygen but is used to describe environments rather than organisms.

archaea (archaebacteria) a large and ancient branch of the bacterial family tree; includes many bacteria that are adapted to life in extreme environments and that have different structures, such as cell walls and flagella, from other bacterial groups. Many specialists put the archaea in a major group separate from other bacteria.

ATP (adenosine triphosphate) an energy-rich molecule used by all organisms to store chemical energy for later use in building cell structures or making chemical reactions go. "Triphosphate" refers to the three phosphates that are part of the molecule; most of the usable energy is stored in the bonds between the phosphates. After energy is released from ATP it becomes ADP (adenosine diphosphate), having lost one phosphate.

aufwuch the community of organisms that live attached to both animate and inanimate surfaces in bodies of water. See also biofilm.

autotroph an organism that makes its own food (sugars) out of carbon dioxide or some other simple starting material. Divided into chemoautotrophs, for which the source of energy is chemical bonds, and photoautotrophs, for which the source of energy is light. Photoautotrophs are often called photosynthesizers.

bacillus both a genus name and a general term for rod-shaped bacteria. May be used as a suffix.

bacterium(a) simple, usually single-celled organism (also called a prokaryote), whose cells do not have subdivided compartments. Some specialists exclude the archaea from the other bacteria, placing them in their own group.

base a chemical with high pH (often bitter), which yields hydroxyl ions (OH⁻) when in water. A sort of chemical opposite to an acid; also called alkaline. Each base has a corresponding salt, indicated by the suffix "-ate" (e.g., sodium bicarbonate).

biofilm a complex microbial community attached to animate or inanimate surfaces. See also aufwuch.

bioluminescence the property of giving off significant light as a byproduct of a chemical reaction. Almost all chemical reactions give off a small amount of light; true bioluminescence occurs in organisms such as fireflies and deep sea animals, dinoflagellates, and bacteria.

biosorption the absorption of chemicals by either living or dead organisms. Many of the chemicals that make up organisms have slight overall negative charges; as a result, they sometimes attract positively charged chemicals such as some metal ions. Some industries use organisms to absorb excess metals, for example in mining wastewaters.

carbon dioxide (CO_2) a gas produced as a waste product of many types of heterotrophic metabolism and used as a starting material in most autotrophic metabolisms. Concentrations are low both in the Earth's atmosphere (about 0.03%) and in water.

chemoautotroph see autotroph.

chlorophyll the green pigment used by photoautotrophs to capture light energy as a first step toward making their own food (sugars).

chlorosis the presence of yellowish pigments when green (chlorophyll) pigments are expected. Often associated with a disease or nutrient deficiency in plants.

coccus(i) a spherical bacterium.

coevolution reciprocal evolutionary change in two or more interacting species.

coliform looking like or related to common rod-shaped bacteria found in intestines, especially *Escherichia coli.*

commensalism a type of symbiosis in which one organism derives food or other benefits from another one without harming it.

consortium a type of symbiotic association in which individuals of two or more different species attach to each other, exchange chemicals, and in general live and reproduce more efficiently in that configuration than as separate individuals.

diatom a single-celled photosynthetic protist (an alga) with an external skeleton of silicon dioxide (glass).

dinoflagellate a single-celled photosynthetic or heterotrophic protist.

DNA (deoxyribonucleic acid) the genetic material of all organisms; in eukaryotes, it is contained within the cell nucleus.

DNA sequencing determining the order of adenosine (A), cytosine (C), thymine (T), and guanine (G) in a DNA molecule. Sequencing is a starting point for constructing phylogenetic trees based on evolutionary changes in DNA.

electron a negatively charged elementary particle.

endolithic living inside rocks, as some bacteria, especially cyanobacteria, and other organisms.

enteric dwelling in the intestines.

eukaryote an organism composed of complex cells that contain compartments, including a nucleus with DNA. Includes humans and all other animals, all plants, all fungi, and all protists.

eutrophic having too much of a particular nutrient—often used for aqueous environments with an abundance of decaying organisms or pollutants, resulting in an overgrowth of some opportunistic organisms. Abundant nutrients often provide perfect conditions for some bacteria, although they may be detrimental to some aerobic species favored by humans.

extremophile loving extreme conditions. Many bacteria thrive in acidic or alkaline habitats and at both ends of the temperature spectrum, habitats considered extreme from a human point of view. See acidophile, alkaliphile, halophile, hyperthermophile, thermophile.

fermentation a chemical reaction by which some organisms break down food molecules to collect the energy stored in the chemical bonds. Fermentation often leaves waste products that themselves constitute food for other organisms with a more elaborate set of chemical reactions, such as respiration.

flagellum(a) a whip-like appendage by which bacteria propel themselves. Flagella are often visible under the microscope only with special mounting and staining procedures, and some bacteria have flagella only in certain life cycle stages or conditions. Thus, the presence or absence of flagella is not a simple diagnostic characteristic for amateurs.

fungi heterotrophic, usually multicellular, eukaryotes, often with thread-like cells (hyphae); includes mushrooms, molds, and the fungal part of lichens.

geyserite a gray-white silica mineral formed in geysers and hot springs.

gleyed characterizing some soils or clays that are devoid of oxidized minerals, especially iron, and thus are gray to blue-gray rather than reddish brown.

Gram stain a purple stain invented by Christian Gram, which is used to divide bacteria into two groups, positive and negative, depending on whether their cell walls accept the stain. Archaea are not classified in either group.

gram-negative bacteria (in many diverse groups) for which the purple Gram stain can be removed by washing with a solvent such as ethanol.

gram-positive bacteria that retain a purple color in their cell walls even after attempts to wash out the Gram stain. The gram-positives constitute a distinct taxonomic group. Most wall-less bacteria are also classified as gram-positive based on their DNA sequences.

gram-variable bacteria that are difficult to Gram stain predictably but turn out to be gram-positive based on the structure of their cell walls.

halophile an organism (mostly archaea) that can grow on salt crystals or in very dense brine (greater than 20% salt). Salt-tolerant organisms can endure salt but do not find it optimal.

heterotroph an organism that uses compounds containing carbon and hydrogen as a source of energy. Any organism, either dead or alive, as well as wastes from organisms, can serve as food for other organisms, as can long-dead organisms in the form of fossil fuels. See also autotroph.

hydrogen the smallest element, present in trace amounts as a gas (H_2) in the Earth's atmosphere or dissolved in water; also a component of water (H_2O), some minerals, and other compounds. It is an essential part of all organisms, and therefore much of the chemical activity of organisms consists of getting, keeping, and using hydrogen.

hypersaline an environment with high salt content, such as concentrated brine or crystalline salt.

hyperthermophile a bacterium for which the optimum temperature is greater than 80°C (176°F). Some hyperthermophiles can grow in temperatures at or above the boiling point of water.

ion an atom with a positive or negative electrical charge. For example, the hydrogen ion (H^+) lacks an electron and therefore has a positive charge, whereas the chlorine ion (Cl^-) has an extra electron and a negative charge. Ions generally participate readily in all sorts of chemical reactions. Positive and negative ions tend to attract each other like the north and south poles of magnets.

Janus Green a stain that indicates the presence of oxidation in an organism.

karst a region characterized by limestone, often riddled with caves and sink holes.

knall gas an explosive mixture of hydrogen and oxygen.

leg hemoglobin a red pigment in root nodules of legumes that binds oxygen and keeps it from interfering with nitrogen fixation.

legume a large group of plants (family Leguminosae) that bear seeds in pods and have symbioses with nitrogen-fixing bacteria in their root nodules.

meiofauna tiny animals (<1 mm) that occupy the spaces between sand grains.

metabolism all of the chemical reactions by which cells acquire and use energy and food, reproduce, and maintain themselves.

methane (CH_4) a gas consisting of one carbon and four hydrogens. It is a trace gas in Earth's atmosphere or dissolved in water. Even in small amounts, it is a "greenhouse" gas. Most methane is probably generated by methanogenic bacteria.

methanogen a group of archaea that produce methane as a waste product of their metabolism.

methanotroph a bacterium (includes those in diverse groups) that uses methane in some aspect of its metabolism.

microfossil an impression or remains of a cell preserved in a sedimentary rock.

micrometer (micron) one thousandth of a millimeter. An average-sized bacterial cell is about 1 micrometer.

nematode a large, diverse phylum of minute worms, also called round worms. Most occupy soils and waters, and a few are parasites.

neuston the community of organisms dwelling in the surface films of water.

nitrogen fixation the metabolic conversion of nitrogen gas (N_2) to a form that organisms can use to build proteins and other essential compounds. Only certain groups of bacteria can perform this essential function; without them, most nitrogen would be unavailable to other organisms.

opportunist an organism with a tendency to grow quickly using a wide variety of abundant resources (weedy). When nutrients are unavailable, opportunists have low activity levels or become dormant. Many pathogens are opportunists that take advantage of breaches in the defenses of potential hosts.

oxidation see reduction.

oxygen a toxic, reactive gas that makes up 20% of Earth's atmosphere and is dissolved in some waters. Oxygen is a major by-product of most types of photosynthesis, the source of most atmospheric oxygen. Organisms have numerous chemical defenses and barriers to prevent oxygen damage, including oxygen respiration, which not only detoxifies oxygen but puts it to use for metabolism.

pathogen an organism that causes disease in other organisms. Only a small minority of bacteria are pathogens.

petri dish a shallow covered dish in which bacteria or other microorganisms are cultured, often on a solid, gel-like nutrient medium.

pH a measure, on a logarithmic scale, of the concentration of hydrogen ions in a solution, ranging in value from 0 to 14. The value for pure water, pH 7, is neutral; the number indicates a concentration of hydrogen ions of 1×10^{-7} grams per liter. Numbers below 7 indicate acidity and numbers above 7 indicate alkalinity: vinegar with pH 3 has 1×10^{-3} grams H/liter, and a sodium bicarbonate solution of pH 8 has 1×10^{-8} grams H/liter.

phase contrast a lighting system for a microscope that makes the cells of microorganisms more visible.

photic zone the layer of water or sediment through which light can pass and therefore photosynthesis can take place.

photoautotroph see autotroph.

photosynthesizer see autotroph.

phylogeny the evolutionary history of an organism or group of organisms. Several techniques are used to decipher phylogenies, including analysis of fossils, changes in DNA sequences, metabolic pathways, and the morphology and other characteristics of organisms. From such studies, evolutionary (phylogenetic) family trees are constructed.

plankton minute animals and plants that float or swim weakly in a body of water.

postgastric located below the stomach (e.g., a digestive chamber).

pregastric located above or associated with the stomach.

prokaryote see bacterium.

prostheca an extension of a bacterial cell containing cytoplasm (cell contents), as distinguished from a stalk, which does not contain cytoplasm.

protist an aquatic, mostly unicellular (or simple multicellular or colonial) eukaryote. Protists' metabolisms are diverse, ranging from photosynthesis to heterotrophy, both aerobic and anaerobic. Examples include paramecia, amoebae, seaweeds, volvox, slime molds, and malaria parasites. The protists make up one of the four eukaryotic kingdoms, the others being the animals, the plants, and the fungi.

proton a positively charged elementary particle equivalent to the nucleus of a hydrogen atom.

pyrite iron sulfide.

reduction the addition of electrons to a compound, thus conferring a negative charge. The opposite of oxidization, which involves losing electrons and becoming positively charged. Oxidation and reduction reactions ("redox reactions") always occur together because the gain and loss of electrons cannot occur without both donors and recipients. Some environments are oxidized (such as the atmosphere) and others are reduced (such as the depths of sulfur-rich sediments).

respiration the metabolic process by which energy is gained from food. Organisms that carry out aerobic respiration require oxygen and produce water and carbon dioxide as waste products; those that carry out anaerobic respiration use compounds such as sulfates or nitrates rather than oxygen and produce different waste products.

rhodophyte a red eukaryotic alga or red seaweed.

ribosome the cell structure where proteins are produced.

rod a rod-shaped bacterium; also called bacillus.

saltern a salt flat where marine water (including that of ancient inland seas) has evaporated, leaving salts behind. Also called salina.

saprophyte a heterotroph that consumes dead organisms or waste products (other than gases) of organisms.

solfataric field a muddy or crusty, hot (often boiling) terrain overlying an area of hot sulfur springs.

spirillum a spiral-shaped bacterium; also used as a suffix.

stromatolite a laminated sedimentary rock made up of the fossil remains of a community of microorganisms, such as cyanobacteria.

sulfuretum(a) an environment rich in sulfurous compounds, especially hydrogen sulfide.

symbiosis an association (often intimate) between two or more organisms of different species such that both are more fit (leave more offspring) in a particular environment than the separate individuals would have been. For example, lichens are symbiotic associations between a fungus and a photosynthesizer (green eukaryotic algae or cyanobacteria or both).

thermophile an organism that inhabits environments with high temperatures (such as compost heaps or hot springs) and grows optimally at temperatures of 45–80°C (113–176°F). See also hyperthermophile.

thiobiota the inhabitants of sulfur-rich environments such as sulfureta.

troglodyte a bacterium or other organism that occupies the spaces between sedimentary particles and the cracks in rocks deep within the Earth. Also called subsurface bacteria.

vibrio a genus name or a general term for bacteria with a comma or S shape. May also be used as a suffix.

wall-less bacteria includes bacteria that lost their walls after becoming internal symbionts and pathogens of other organisms, as well as many archaea. The former are usually classified as gram-positives based on DNA sequences, whereas the latter are not classified according to the Gram stain.

wet mount a preparation in which a small sample of bacteria or other microscopic material is placed on a glass microscope slide with a drop of water and a thin glass piece (coverslip) placed on top to flatten the drop and protect the microscopic lens.

wet wood (slime flux) a nonpathological condition of some trees caused by methanogens. Methane and other bacterial products sometimes cause fissures in the bark through which they leak.

Winogradsky column a method developed by the Russian microbiologist Sergei Winogradsky for culturing a mixed community of bacteria so that the macroscopic field marks are visible.

References

Ahmadjian, Vernon, and Surindar Paracer. *Symbiosis: An Introduction to Biological Associations.* 1st ed. Hanover, N.H.: University Press of New England, 1986.

Alexander, Martin. *Introduction to Soil Microbiology.* 2d ed. New York: Wiley, 1977.

The American Society for Microbiology (ASM). http://www.asmusa.org.

Annual Review of Microbiology. Stanford, Calif.: Annual Reviews, 1947–.

Astrobiology Micro*scope, The Astrobiology Institute, Marine Biological Laboratory (MBL), Woods Hole.
  http://www.mbl.edu/baypaul/microscope/general/page_01b.htm.

Balows, A., H. Trüper, M. Dworkin, W. Harder, and K.-H. Schleifer, eds. *The Prokaryotes: A Handbook on the Biology of Bacteria: Ecophysiology, Isolation, Identification, Applications.* 2d ed. 4 vols. New York: Springer-Verlag, 1992.

Biddle, Wayne. *A Field Guide to Germs.* 1st ed. New York: Henry Holt, 1995.

Black, Jacquelyn G. *Microbiology: Principles and Applications.* 3d ed. Upper Saddle River, N.J.: Prentice Hall, 1996.

Brock, T. D., M. Madigan, J. Martinko, and J. Parker. *Biology of Microorganisms.* 7th ed. Englewood Cliffs, N.J.: Prentice Hall, 1994.

Brodo, Irwin M., Sylvia Duran Sharnoff, and Stephen Sharnoff. *Lichens of North America.* New Haven, Conn.: Yale University Press, 2001.

Brown, Sanborn C. *Wines and Beers of Old New England: A How-to-Do-It History.* Rev. ed. Hanover, N.H.: University Press of New England, 1998.

Canter-Lund, Hilda, and J. W. G. Lund. *Freshwater Algae: Their Microscopic World Explored.* Bristol, Engl.: Biopress Ltd., 1995.

Darwin, Charles. *Voyage of the Beagle.* 1860. Garden City, N.Y.: Doubleday, 1962.

References

Dekkers, Midas. *The Way of All Flesh: A Celebration of Decay.* Translated by Sherry Marx-Macdonald. New York: Farrar, Straus and Giroux, 2000.

De Kruif, Paul. *The Microbe Hunters.* New York: Harcourt, Brace, 1927.

Dixon, Bernard. *Magnificent Microbes.* New York: Athaneum, 1976.

Dixon, Bernard. *Power Unseen: How Microbes Rule the World.* New York: W. H. Freeman, 1994.

Dobell, Clifford. *Antony van Leeuwenhoek and His "Little Animals": Being Some Account of the Father of Protozoology and Bacteriology and His Multifarious Discoveries in These Disciplines.* New York: Russell and Russell, 1958.

Dyer, Betsey Dexter, and Robert Alan Obar. *Tracing the History of Eukaryotic Cells: The Enigmatic Smile.* New York: Columbia University Press, 1994.

Fenchel, Tom, and Bland J. Finlay. *Ecology and Evolution in Anoxic Worlds.* New York: Oxford University Press, 1995.

Fenchel, Tom, G. M. King, and T. H. Blackburn. *Bacterial Biogeochemistry: The Ecophysiology of Mineral Cycling.* San Diego: Academic, 2000.

Fortey, Richard. *Trilobite! Eyewitness to Evolution.* New York: Knopf, 2000.

Guinard, Jean-Xavier. *Lambic.* Boulder, Colo.: Brewers, 1990.

Holt, John G., N. Krieg, P. Sneath, J. Staley, and S. Williams. *Bergey's Manual of Determinative Bacteriology.* 9th ed. Baltimore: Williams and Wilkins, 1994.

Holt, John G., et al., eds. *Bergey's Manual of Systematic Bacteriology.* 4 vols. Baltimore: Williams and Wilkins, 1984–89.

Howland, John L. *The Surprising Archaea: Discovering Another Domain of Life.* New York: Oxford University Press, 2000.

Krumbein, Wolfgang E., D. Paterson, and L. Stal, eds. *Biostabilization of Sediments.* Oldenburg: Bibliotheks- und Informationssystem der Univ., 1994.

Levine, Joseph. "The Immortal Thread." Directed and produced by James F. Galway. In *The Secret of Life.* WGBH-Boston, 1993. Videocassette. In print as "The Immortal Thread," in Joseph Levine and David Suzuki, *The Secret of Life: Redesigning the Living World,* Boston: WGBH Boston, 1993.

Luard, Elisabeth. *The Old World Kitchen.* North Pomfret, Vt.: Trafalgar Square, 2000.

Margulis, Lynn, and Michael F. Dolan. *Early Life: Evolution on the Precambrian Earth.* 2d ed. Sudbury, Mass.: Jones and Bartlett, 2002.

Margulis, Lynn, and Karlene V. Schwartz. *The Five Kingdoms: An Illustrated Guide to the Phyla of Life on Earth.* 3d ed. New York: Freeman, 1998.

Marples, Mary J. *The Ecology of the Human Skin.* Springfield, Ill.: Thomas, 1965.

Mayle, Peter. *French Lessons: Adventures with Knife, Fork, and Corkscrew.* New York: Vintage, 2002.

Moore, Peter D. *Wetlands.* New York: Facts on File, 2000.

The National Center for Biological Information. GenBank, molecular databases, and literature databases. http://www.ncbi.nlm.nih.gov.

Needham, Cynthia A., Mahlon Hoaglan, Bert Dodson, and Kenneth McPherson. *Intimate Strangers: Unseen Life on Earth.* Washington, D.C.: ASM Press, 2000. Video series, *Intimate Strangers: Unseen Life on Earth*, produced by Baker and Simon Associates, in association with Oregon Public Broadcasting and the American Society for Microbiology, 4 hrs., 1999, videocassette.

Nelson, Dick, and Sharon Nelson. *Hiker's Guide to Glacier National Park.* Glenwood, N.M.: Tecolote, Glacier History Association, 1978.

Noble, W. C., and Dorothy A. Somerville. *Microbiology of Human Skin.* Vol. 2 of *Major Problems in Dermatology.* London: W. B. Saunders, 1974.

Pliny the Elder. *Natural History.* 10 vols. Loeb Classical Library. Cambridge: Harvard University Press, 1938–63.

Postgate, John. *Microbes and Man.* 3d ed. Cambridge: Cambridge University Press, 1992.

Postgate, John. *The Outer Reaches of Life.* Cambridge: Cambridge University Press, 1994.

Rossmore, Harold W. *The Microbes, Our Unseen Friends.* Detroit: Wayne State University Press, 1976.

Roueche, Berton. *The Medical Detectives.* 2 vols. New York: Times Books, 1980–84.

Sagan, Dorion, and Lynn Margulis. *Garden of Microbial Delights.* 1st ed. Boston: Harcourt, Brace, Jovanovich, 1988.

Schreier, Carl. *A Field Guide to Yellowstone's Geysers, Hot Springs, and Fumaroles.* Moose, Wyo.: Homestead, 1992.

Shephard, Sue. *Pickled, Potted, and Canned: How the Art and Science of Food Preserving Changed the World.* New York: Simon and Schuster, 2001.

Sonea, Sorin, and Maurice Panisset. *A New Bacteriology.* Boston: Jones and Bartlett, 1983.

References

Stanier, Roger Y., Michael Doudoroff, and Edward A. Adelberg. *The Microbial World*. 2d ed. Englewood Cliffs, N.J.: Prentice Hall, 1963.

Steingarten, Jeffrey. *The Man Who Ate Everything: And Other Gastronomic Feats, Disputes, and Pleasurable Pursuits*. New York: Knopf, 1997.

Tannock, G. W. *Normal Microflora: An Introduction to Microbes Inhabiting the Human Body*. New York: Chapman and Hall, 1995.

Thomas, Lewis. *Lives of a Cell: Notes of a Biology Teacher*. New York: Viking, 1974.

Visser, Margaret. *Much Depends on Dinner: The Extraordinary History and Mythology, Allure and Obsessions, Perils and Taboos, of an Ordinary Meal*. 1st ed. New York: Grove, 1987.

Zook, Douglas, et al. *The Microcosmos Curriculum Guide to Exploring Microbial Space*. Dubuque, Iowa: Kendall/Hunt, 1992.

Index

Index

Index

Index

Index

Index